Tao Classic
of
Longevity
and Immortality

Sacred Wisdom
and Practical Techniques

DR. & MASTER ZHI GANG SHA

BenBella Books, Inc.
Dallas, TX

Copyright © 2018 Heaven's Library Publication Corp.
heavenslibrary@drsha.com

BenBella Books, Inc.
10440 N. Central Expressway, Suite 800
Dallas, TX 75231
www.benbellabooks.com
Send feedback to feedback@benbellabooks.com.

The Heaven's Library logo is a trademark of Heaven's Library Publication Corp.

Printed in the United States of America
10 9 8 7 6 5 4 3 2 1

Library of Congress Cataloging-in-Publication Data is available upon request.
ISBN 9781946885531
e-ISBN 9781946885586

Editing by Leah Wilson
Copyediting by Stacia Seaman
Proofreading by Cape Cod Compositors, Inc.

Text design and composition by Aaron Edmiston
Cover design by Henderson Ong

Distributed to the trade by Two Rivers Distribution, an Ingram brand
www.tworiversdistribution.com

Special discounts for bulk sales (minimum of 25 copies) are available.
Please contact Aida Herrera at aida@benbellabooks.com.

Contents

List of Figures

Introduction

A HUMAN BEING has two parts of life, a physical life and a spiritual life. Most people are focusing on their physical life. Many people are aiming on their spiritual life. Ancient Chinese wisdom teaches that everyone and everything is made of Shen Qi Jing (神氣精). "Shen" includes *soul, heart, and mind.* "Qi" means *energy.* "Jing" means *matter.* Shen Qi Jing includes the physical life and the spiritual life. Therefore, both physical life and spiritual life are very important. In this book, *Tao Classic of Longevity and Immortality: Sacred Wisdom and Practical Techniques,* I intend to serve humanity with sacred Tao wisdom and practical techniques to balance and develop both lives. Eventually, human beings can reach their greatest potentials in both lives: longevity, immortality, and attaining Tao.

What are the most important things in human life? You would agree with me that the most valuable things for a human being are good health and happiness. Almost every human being would love to have a long life. However, longevity is meaningless without quality. The quality of physical life depends on the evolution of the spiritual life.

Immortality is the highest achievement for one's physical and spiritual journeys. Immortality is hard to believe because it is simply not easy to encounter an immortal. There are legends of some Taoist saints, Buddhas, and other saints in history who lived for hundreds or thousands of years. Most of us have not seen these kinds of saints or Buddhas. Therefore, immortality seems just to be in legends or in realms of fantasy.

Yet in this book I will tell you otherwise. I will share Tao secrets, wisdom, knowledge, and practical techniques that make longevity and immortality possible and achievable. If you have already read my previous *New York Times* bestsellers, *Tao I*[1] and *Tao II,*[2] you would agree with me that Tao Normal Creation and Tao Reverse Creation are the most profound wisdom, philosophy, and science of all lives, including countless planets, stars, galaxies,

---

1 Dr. and Master Zhi Gang Sha, *Tao I: The Way of All Life,* New York/Toronto: Atria Books/Heaven's Library, 2010.
2 Dr. and Master Zhi Gang Sha, *Tao II: The Way of Healing, Rejuvenation, Longevity, and Immortality,* New York/Toronto: Atria Books/Heaven's Library, 2010.

and universes. These two creations have unlimited levels of secrets, wisdom, knowledge, and practical techniques to help you reach longevity, immortality, and an even higher spiritual achievement: reach Tao. However, it is a lifelong or even many-lives-long study and practice to comprehend and achieve immortality and reach Tao. If you determine to pursue this, you would receive more and more "aha" or "wow" realization moments on this journey.

First, I would like to share wisdom from the foremost wise man Lao Zi (or Lao Tzu), the author of *Dao De Jing*[3] (or *Tao Te Ching*). *Dao De Jing* is a fundamental text for both philosophy and religion. It is the ancient classic to learn Tao wisdom and practice. It has been studied and practiced by philosophers and Taoists in China for more than two thousand years. It is ancient wisdom for modern times.

Tao Normal Creation

Dao Sheng Yi (道生一)

Dao (Tao) is the Source. Tao is the Ultimate Creator. "Sheng" means *create*. "Yi" means *oneness*. Oneness is named Hun Dun Yi Qi (渾沌一氣). "Hun Dun" means *blurred, formless*. "Yi" means *oneness*. "Qi" means *energy*. "Hun Dun Yi Qi" means *the oneness of blurred energy*. "Dao Sheng Yi" means *Tao creates Hun Dun Yi Qi, the Oneness condition*. See figure 1.

Yi Sheng Er (一生二)

"Er" means *two*. "Yi Sheng Er" means *Oneness creates Two: Heaven and Earth*. Within the oneness of blurred energy there is the mixture of two types of Qi (energy): Qing Qi (清氣) and Zhuo Qi (濁氣). Qing Qi is clean and light energy. Zhuo Qi is unclean or heavy energy. This Oneness condition has waited for Qi Hua (氣化), which is the transformation of Qi. When Qi Hua occurs, Qing Qi rises to form Heaven. Zhuo Qi falls to form Mother Earth. Heaven is the physical manifestation of Yang. Mother Earth is the physical

---

3 An interesting modern commentary is Wang Keping, *The Classic of the Dao: A New Investigation*, Beijing: Foreign Languages Press, 1998.

manifestation of Yin. Yin and Yang are parts of the oneness of Tao. A human being lives between Heaven and Earth. Therefore, a human being lives within the Yin Yang world.

Er Sheng San (二生三)

"San" means *three*. "Er Sheng San" means *Two creates Three: Oneness, Heaven, and Mother Earth.*

San Sheng Wan Wu (三生萬物)

"Wan" means *ten thousand*. In Chinese, "ten thousand" represents *innumerable*. "Wu" means *thing*. "San Sheng Wan Wu" means *Three creates everything*. Oneness plus Heaven and Mother Earth create countless planets, stars, galaxies, and universes and everything.

I called the process described in these four phrases of Lao Zi "Tao Normal Creation" and explained it in detail in my *Tao I* and *Tao II* books. Tao and One belong to the Wu World. This "Wu" means *emptiness* or *formlessness*, which is the Creator and Ultimate Source. Two, Three, and Wan Wu belong to the You World. "You" means *existence with forms*.

How are human beings and countless planets, stars, galaxies, and universes created? The Wu World creates the You World. This is the answer that billions of people have searched for since creation. This is the great answer from the greatest wisdom man, Lao Zi.

Lao Zi also explained the immortality path in *Dao De Jing* through the phrases "Ren Fa Di, Di Fa Tian, Tian Fa Dao, Dao Fa Zi Ran (人法地, 地法天, 天法道, 道法自然)."

Ren Fa Di (人法地)

There is an ancient statement, "Ren Wei Wan Wu Zhi Ling (人為萬物之靈)." "Ren" means *human being*. "Wei" means *is*. "Wan Wu" means *all things, including countless planets, stars, galaxies, and universes*. "Zhi" means *their*. "Ling" means *soul, spirit*. "Ren Wei Wan Wu Zhi Ling" means *human beings are the leading soul of all things.* "Fa" means *follow.* "Di" means *Mother Earth.* "Ren Fa Di" means *human beings follow the principles and laws of Mother Earth.* For example, there is a weather law on Mother Earth. In the summer it is very hot in some parts of Mother Earth. People need to dress in summer clothes. If you were to wear winter clothes during summer heat, you could get sick or even die. In one sentence: *Human beings must follow the laws of Mother Earth.* This is Ren Fa Di.

Di Fa Tian (地法天)

"Di" means *Mother Earth.* "Fa" means *follow.* "Tian" means *Heaven.* "Di Fa Tian" means *Mother Earth follows the principles and laws of Heaven.*

The Shen Qi Jing of Heaven is purer than the Shen Qi Jing of Mother Earth. The sacred wisdom of Di Fa Tian is that after a human being's Shen Qi Jing transforms to Mother Earth's Shen Qi Jing, it then further transforms to Heaven's Shen Qi Jing. This is the second step of immortality. This is another level of sacred Tao wisdom and practice that Lao Zi revealed to humanity.

Tian Fa Dao (天法道)

Dao (Tao) is the Source and Ultimate Creator. "Tian" belongs to the You World, which is the existence world with forms. Tao is the Wu World. "Tian Fa Dao" means *from the You World return to the Wu World.* This is the third step of immortality. It further transforms Heaven's Shen Qi Jing to Tao's Shen Qi Jing. As the Source and Ultimate Creator, Tao has the purest Shen Qi Jing.

Dao Fa Zi Ran (道法自然)

Dao is the Source. In this phrase, "Fa" can be interpreted as *yes.* "Zi Ran" means *nature.* "Dao Fa Zi Ran" means *Tao is nature* or *follow nature's way.* Dao Fa Zi Ran is the state of immortality. Whatever your action, whatever your behavior, whatever your speech, and whatever your thought—all are completely melding with nature or completely following all of the universal principles and laws. Immortality is to live forever. This is the highest achievement for one's physical life as well as for one's spiritual life. Immortality is the physical life and spiritual life joining as one. It is very difficult to believe in or even comprehend immortality if you do not have sacred spiritual wisdom. In *Dao De Jing*, Lao Zi released this profound sacred wisdom together with practice methods to reach immortality.

Figure 1. Tao Normal Creation and Tao Reverse Creation

Although *Dao De Jing* is the most captivating masterpiece, it is the most difficult to comprehend. Many people have spent their entire lives dedicated to the study of *Dao De Jing* to limited benefit, just to gain a basic understanding.

Let me explain in a simple way. I called the immortality pathway "Tao Reverse Creation." As I mentioned earlier, all things are made of Shen Qi Jing. A human being is made of Shen Qi Jing. Mother Earth is made of Shen Qi Jing. Heaven is made of Shen Qi Jing. Tao is the Creator of Shen Qi Jing. However, there are enormous differences in the quality, frequency, and vibration of the Shen Qi Jing among them. In order to achieve longevity and immortality and then return to Tao, human beings must transform their Shen Qi Jing to Mother Earth's, Heaven's, and Tao's Shen Qi Jing to reverse the Tao Normal Creation process.

Tao Reverse Creation

The immortality path of Tao Reverse Creation can be explained through the phrases "Wan Wu Gui San, San Gui Er, Er Gui Yi, Yi Gui Dao (萬物歸三, 三歸二, 二歸一, 一歸道)."

Wan Wu Gui San (萬物歸三)

"Gui" means *return*. A human being is in the Wan Wu layer of the You World. Tao wisdom, knowledge, and messages are encoded in the human body. When you do Xiu Lian (修煉) practice, you will unlock the code to connect with Tao, in order to return to Tao. "Xiu" means *purification*. "Lian" means *practice*. "Xiu Lian" is *purification practice to purify our Shen Qi Jing to the next level*. The Xiu Lian process takes us through our entire spiritual journey and physical journey.

"Wan Wu Gui San" means *all things return to Three*. Countless planets, stars, galaxies, and universes, including human beings, return to Three, which is Hun Dun Yi Qi plus Heaven and Mother Earth. It is the reverse of San Sheng Wan Wu, *Three creates everything*, in Tao Normal Creation.

San Gui Er (三歸二)

"San Gui Er" means *Three returns to Two*. Hun Dun Yi Qi, Heaven, and Mother Earth return to Heaven and Mother Earth. It is the reverse of Er Sheng San, *Two creates Three*.

Er Gui Yi (二歸一)

"Er Gui Yi" means *Two returns to One*. The Shen Qi Jing of Heaven and Mother Earth join as one to go back to Oneness. It is the reverse of Yi Sheng Er, *One creates Two*.

Yi Gui Dao (一歸道)

"Yi Gui Dao" means *the Shen Qi Jing of the Hun Dun Yi Qi condition goes back to Tao Source.* It is the reverse of Dao Sheng Yi, *Tao creates Hun Dun Yi Qi, the Oneness condition.*

You may have noticed that my books and my teaching have become more and more simple. They directly come to the point. In Chinese, it is called Da Dao Zhi Jian (大道至簡). "Da" means *big.* "Dao" means *the Way.* "Zhi" means *extremely.* "Jian" means *simple.* "Da Dao Zhi Jian" means *the big Way is extremely simple.* I hope the simplicity of my teaching can make your longevity and immortality journey simpler as well.

The longevity and immortality journey is everyone's spiritual journey and physical journey. There is a renowned saying by Confucius: "Dao Bu Yuan Ren, Ren Zi Yuan (道不遠人，人自遠)," which means *Tao (the Way) is not far from a human being. A human being is far from Tao.* Tao is in a human being and in every aspect of our lives. Tao is the universal principles and laws. However, a human being is far from Tao because a person's Shen Qi Jing blockages prevent that person from following the universal laws and principles of Tao.

Every aspect of life also has its Xiao Dao (小道, *small way and principles*). The whole universe has its Da Dao (大道, *big way and principles*). Why can people not see Tao? Because of Shen Qi Jing blockages. Shen includes soul, heart, and mind. Soul blockages include all kinds of negative karma. Heart blockages include greed (貪, tan), anger (嗔, chen), and lack of wisdom (癡, chi). Mind blockages include negative mind-sets, negative beliefs, negative attitudes, ego, attachments, and more. Qi blockages are energy blockages, which occur between the cells. Jing blockages are matter blockages, which occur inside the cells. In one sentence, Shen Qi Jing blockages are the pollution in one's spiritual journey and physical journey. On the Tao journey, the first step is to remove Shen Qi Jing blockages.

In summary, longevity and immortality are possible and achievable. However, serious purification and practice are musts. All of the sacred wisdom and practical techniques presented in this book will empower you on your Tao journey.

Open your heart and soul.
Learn the sacred wisdom of longevity and immortality.
Purify your Shen Qi Jing, which is your soul, heart, mind, and body.

Practice. Practice. Practice.
Apply the secrets, wisdom, knowledge, and practical techniques in every
*   aspect of your life.*
Enlighten your soul, heart, mind, and body.
Follow the path of Tao Reverse Creation: Wan Wu Gui San, San Gui Er,
*   Er Gui Yi, Yi Gui Dao, Dao Fa Zi Ran or Ren Fa Di, Di Fa Tian,*
*   Tian Fa Dao, Dao Fa Zi Ran.*
Finally, you could reach longevity and immortality.

1. Tao Source sacred wisdom

To understand the wisdom, knowledge, and practical techniques contained within the new Tao Classic that is the core of this book, it is first essential to understand some important Tao terms and fundamental knowledge.

What is Tao?

Tao is the Source. Tao is the Ultimate Creator. Tao is immortality. Tao is the universal laws, rules, and principles. In *Dao De Jing*, Lao Zi explained that Tao is the blurred condition. Within the blurred condition there is Shen Qi Jing. The Shen Qi Jing of Tao is the purest and formless Shen Qi Jing.

The nature of Tao

The nature of Tao consists of ten essential qualities that are required for any serious Tao practitioner. I have explained these ten greatest qualities in detail in my book with Adam Markel, *Soul Over Matter*.[4]

---

4 Dr. and Master Zhi Gang Sha and Adam Markel, *Soul Over Matter: Ancient and Modern Wisdom and Practical Techniques to Create Unlimited Abundance*, Dallas/Toronto: BenBella Books/Heaven's Library, 2016.

Da Ai (大愛)	greatest love
Da Kuan Shu (大寬恕)	greatest forgiveness
Da Ci Bei (大慈悲)	greatest compassion
Da Guang Ming (大光明)	greatest light
Da Qian Bei (大謙卑)	greatest humility
Da He Xie (大和諧)	greatest harmony
Da Chang Sheng (大昌盛)	greatest flourishing
Da Gan En (大感恩)	greatest gratitude
Da Fu Wu (大服務)	greatest service
Da Yuan Man (大圓滿)	greatest enlightenment

Why does a person need to study and reach Tao?

Millions of people wish to have good health and happiness. Millions of people wish to have longevity. Millions of people are dreaming of immortality. Tao is immortality. Tao is the Source and Ultimate Creator. To study and practice Tao is to reach good health, happiness, longevity, and immortality.

Xiu Xian Cheng Dao (修仙成道) is an ancient wisdom and practice since creation. "Xiu" means *purification*, including purification of Shen Qi Jing. "Xian" means *saint*. "Cheng" means *to reach*. "Dao" means *immortality, the Source, or Ultimate Creator*. "Xiu Xian Cheng Dao" is to *purify your Shen Qi Jing to become a saint in order to finally reach Tao, which is immortality*. To study and practice Tao can transform all our life.

All life includes:

- boosting energy, stamina, vitality, and immunity
- healing the spiritual, mental, emotional, and physical bodies
- transforming all kinds of relationships
- transforming finances
- opening spiritual channels
- increasing wisdom and intelligence
- enlightening the soul, heart, mind, and body
- bringing success in every aspect of life
- prolonging life and moving toward immortality.

Wisdom of karma

To reach longevity and immortality, the first step is to heal our spiritual, mental, emotional, and physical bodies. I cannot emphasize enough the ancient wisdom that everyone and everything is made of Shen Qi Jing. Sickness is due to Shen Qi Jing blockages. These blockages result directly from one's negative karma.

What is karma, then? Karma is the record of one's services in all lifetimes, including this lifetime and past lifetimes. Karma includes positive karma and negative karma. Positive karma means that a person and that person's ancestors have offered love, care, compassion, sincerity, honesty, generosity, kindness, and more to others. Positive karma brings good health, happiness, blessed relationships, flourishing finances, and a good life.

Negative karma is the mistakes a person and that person's ancestors have made in this lifetime and past lifetimes, including killing, harming, taking advantage of others, cheating, stealing, and more. Negative karma is the root cause of sickness, relationship and financial challenges, lack of wisdom, and failure in any aspect of life. For instance, if you suffer knee pain in this life, you and your ancestors might have hurt others' knees in previous lifetimes or even in this lifetime. If you are depressed, you and your ancestors might have caused others to experience depression.

In one sentence:

**What you are suffering in the spiritual, mental,
emotional, and physical bodies is due to the suffering
that you and your ancestors have caused to others in their
spiritual, mental, emotional, and physical bodies.**

Billions of people worldwide believe in karma. Billions of people have not yet realized this sacred wisdom and truth. I am delighted to share the truth of this sacred wisdom and practice.

In one sentence:

**Karma is the root cause of success and
failure in every aspect of life.**

Shen includes soul, heart, and mind. Soul blockages are negative karma. Because ancient wisdom teaches that the heart houses the mind and soul, it is vital to remove heart blockages for healing all life. Heart blockages include impurities, selfishness, greed, anger, and lack of wisdom in actions, behaviors, speech, and thoughts. Mind blockages include negative mind-sets, negative beliefs, negative attitudes, ego, and attachments. Ego is one of the greatest blockages in life. A renowned spiritual father in history, Wang Yang Ming (王陽明), said, "Ego is the biggest enemy in one's spiritual journey and physical journey." To purify the ego, humility is the answer. Greatest humility is the fifth of the Ten Da natures bestowed by Tao Source.

Three key steps to accomplish longevity and immortality

In ancient wisdom, all sickness in the spiritual body, mental body, emotional body, and physical body is caused by "lou ti (漏體)." "Lou" means *leak*. "Ti" means *body*. A leaky human body is like a cup with cracks. Fill the cup with water and the water gradually leaks out. A leaky human body is an unhealthy body that gradually leaks its Shen Qi Jing. To heal all sickness, the leaky body must be repaired, transformed, and filled.

A human being's body has two major parts: spaces and organs. The major spaces include:

- seven Soul Houses
- San Jiao: the pathway of Qi and bodily fluid
- Wai Jiao: the biggest space in the body, located in front of the spinal column from the neck to the tailbone, plus the space in the back of the head
- Jing Channel
- Qi Channel
- Shen Channel.

Step One: Bu Lou Zhu Ji (補漏築基), fulfill the leaky body and build a foundation

"Bu" means *fulfill*. "Lou" means *leak*, referring to the body. "Zhu" means *to build*. "Ji" means *foundation*. "Bu Lou Zhu Ji" means *fulfill the leaky body and build a foundation*. It includes healing all sickness in the spiritual, mental, emotional, and physical bodies by removing all layers of Shen Qi Jing blockages. To study and chant the new Tao Classic is to remove Shen Qi Jing blockages in the spaces and organs. You are actually repairing and fulfilling your leaky body in the spaces and organs simultaneously. You are purifying your spaces and organs.

Step Two: Ming Xin Jian Xing (明心見性), enlighten the heart to see one's own true nature, Yuan Shen (元神), which is one's Tao nature

Ming Xin Jian Xing is the second major step to reach longevity and immortality. "Ming" means *enlighten*. "Xin" means *heart*. "Jian" means *to see*. "Xing" means *nature of Tao, the Divine, Buddha, and saints*. "Ming Xin Jian Xing" means *enlighten the heart to see one's own true nature*. It is the pathway that one must go through to accomplish longevity and immortality.

Enlighten your heart to see your Yuan Shen, your Tao nature. Every human being has his or her true nature, called Yuan Shen (元神). It is the nature of Tao, the nature of Buddha, the nature of saints in all realms.

Why can't most people see their own true nature? It is because of Shen Qi Jing blockages. Ming Xin is to clear Shen Qi Jing blockages, and then Jian Xing is to have a realization of one's Tao nature. Everyone has both a Shi Shen (識神), one's reincarnated soul, and a Yuan Shen, the true nature of one's self. Yuan Shen is hidden in the area of the Ming Men acupuncture point (located on the lower back directly behind the navel). Shi Shen is located in the heart and is in charge of one's life. If one reaches Ming Xin, one has removed many Shen Qi Jing blockages. One can then see the Yuan Shen. Yuan Shen is the true boss of one's life. In ancient wisdom, Ming Xin Jian Xing is to reach the saints' level. To meld Shi Shen with Yuan Shen is to reach Tao. It is the highest achievement of Tao practice.

Many saints are recognized by the various religions. Saints have enlightened souls, hearts, minds, and bodies. There are countless Buddhas, holy saints, Taoist saints, lamas, gurus, kahunas, and more. They reside in the fourth dimension. Human beings are in the third dimension. A human's Shen Qi Jing has much lower frequency, vibration, and quality than a saint's. So, to become a saint, one has to learn from the saints' example: offer service to others to make others healthier and happier. When one serves, Heaven will uplift one's soul, heart, mind, and body according to the level of the service. Service can be given in every aspect of life through Shen Kou Yi (身口意). "Shen" means *action*. "Kou" means *speech*. "Yi" means *thought*. We have shared the Ten Da qualities above. They are the greatest natures of Tao, the greatest principles to follow in action, speech, and thought to provide the greatest service.

Step Three: Bao Yuan Shou Yi, Shen Qi Jing He Yi (抱元守一, 神氣精合一), hold, focus, and meld with Tao, soul heart mind body join as one

Bao Yuan Shou Yi is the final step to reach Tao, to become a Buddha or a highest saint in other spiritual realms. They are the same. This is the highest achievement in one's spiritual journey and physical life. What is Bao Yuan Shou Yi? "Bao" means *hold*. "Yuan" means *original*, and specifically *Yuan Shen, Yuan Qi, and Yuan Jing*, which are the Shen Qi Jing of Tao. "Shou" means *focus*. "Yi" means *Oneness*. Tao creates Oneness. "Bao Yuan Shou Yi" means *hold and focus on your Yuan Shen, Yuan Qi, and Yuan Jing and meld with Tao*.

Let's use Tao Normal Creation and Tao Reverse Creation in figure 1 to explain this process.

Tao Normal Creation is Tao → One → Two → Three → Wan Wu.

Tao Reverse Creation is Wan Wu → Three → Two → One → Tao.

I would like to remind you of the sacred wisdom that Heaven and Mother Earth belong to the You (existence) World. The number one law in the You World is Yin Yang law. The You World is the Yin Yang world.

Tao and Oneness are the Wu (emptiness, formless) World. In the Wu World, the law is Bao Yuan Shou Yi, Oneness. If you are in the Yin Yang world, you have not reached Tao or become immortal because of the dual nature of Yin and Yang. To reach Tao and become immortal, you must apply the immortal law. It is the Wu World law, Bao Yuan Shou Yi, to become Oneness.

Zhong (中): the core of life

The Chinese character 中 (*zhong*) represents the core of life. For a human being, it is the fundamental area for Tao practice. It is located at "Kun Gong Ming Men Wei Lü Hui Yin Zu Gen Zhong (坤宮命門尾閭會陰足跟中)," which means *Kun Temple, Ming Men acupuncture point, sacrum, perineum, plus the heels of the feet.* Everyone and everything has a Zhong. A human's Zhong is Ren Zhong (人中). Mother Earth's Zhong is Di Zhong (地中). Heaven's Zhong is Tian Zhong (天中). Tao's Zhong is Dao Zhong (道中).

Zhong is the number one area for all kinds of spiritual practitioners to reach Tao, because the core of a human being, Mother Earth, Heaven, and Tao can be joined as one in one's Zhong.

In chapter five of *Dao De Jing*, Lao Zi revealed a sacred statement for the Tao practice of Zhong: "Duo Yan Shu Qiong, Bu Ru Shou Zhong (多言數窮, 不如守中)." "Duo" means *much.* "Yan" means *talk.* "Shu Qiong" means *end.* "Bu Ru" means *not as good as.* "Shou" means *focus.* "Zhong" means *core of life.* "Duo Yan Shu Qiong, Bu Ru Shou Zhong" means *endless talking is not as good as focusing on the Zhong.* This sacred teaching and practice has been the key guidance for humanity to do Tao practice to achieve longevity and immortality.

2. Practice

Studying Tao wisdom and doing Tao practices are the Yin and Yang to achieve longevity and immortality. To only study the wisdom without practice will not yield any results. I teach and emphasize the Four Power Techniques to lead you to do Tao practices chapter by chapter.

The Four Power Techniques are:

Body Power. Body Power is to use hand and body positions to purify and to remove Shen Qi Jing blockages in order to heal, rejuvenate, prolong life, and move forward on the pathway of immortality. One example is to grip your left thumb with your right hand effortlessly. Then place both hands on your lower abdomen, below the navel.

Soul Power. Soul Power⁵ is to say *hello* to the inner souls of your body, including the souls of the spaces, systems, organs, cells, cell units, DNA, and RNA, as well as to outer souls, including saints, healing angels, archangels, ascended masters, lamas, gurus, all kinds of spiritual fathers and mothers, the Divine, and Tao Source, to invoke their power for healing, rejuvenation, and life transformation.

Mind Power. Mind Power is creative visualization. The most important wisdom is to visualize golden or rainbow light radiating in your body to transform your health, relationships, finances, and more.

Sound Power. Sound Power is to chant or sing the new Tao Classic. You can follow me to chant the new Tao Classic accompanied by beautiful Tao music. The recording will guide you in the proper pronunciations and the best melody for each of the one hundred forty-six lines. My singing will bless and empower your longevity, immortality, and Tao journey every time you listen or chant to it. You can purchase it for download here: http://webshop-ca.drsha.com/melding-with-tao-classic-of-longevity-immortality.html.

3. How to use this book

This book has three parts and nineteen chapters:

> Part I: Tao Natures
> > In eleven chapters, you will learn how to use the Ten Da Tao Natures to heal and to develop and enlighten the key spaces in the body.
>
> Part II: Tao Body
> > In two chapters, you will learn further how to heal and how to develop and enlighten the organs of your body.
>
> Part III: Attain Tao
> > In six chapters, you will learn how to join your Shen Qi Jing as one and meld with Tao. Teachings include:

---

5 Dr. and Master Zhi Gang Sha, *The Power of Soul: The Way to Heal, Transform, and Enlighten All Life,* New York/Toronto: Atria Books/Heaven's Library, 2009.

- how to form and grow the Jin Dan until it reaches the size of your body
- the state of achieving immortality and reaching Tao.

Each chapter contains three sections:

1. text of Tao Classic in Chinese and Pinyin with English translation
2. sacred wisdom of the Tao Classic
3. practice: Use the Four Power Techniques to practice the sacred wisdom.

Figures are provided to guide the practice step by step. Note that the chanting in each chapter always starts from the beginning of the Tao Classic, because it is important to activate each Soul House before starting to learn the new practice. In this way, you can keep your practice consistent to receive optimum benefits.

To help you practice the entire Tao Classic, you will find the complete text in the appendix at the end of this book.

Finally, I strongly suggest that you study my *Tao I* and *Tao II* books before you continue in this book. Those two earlier books will help you build fundamental knowledge and prepare your body with basic Tao training. In that way, you can get much better results when you study this book.

I wish you a joyful and successful longevity, immortality, and Tao journey!

Zhi Gang Sha

Part I
Tao Natures

IN PART I, I will teach you how to use the Ten Da Tao Natures to heal, develop, and enlighten the major spaces in your body, including the seven Soul Houses, San Jiao, Wai Jiao, Ming Men, Lower Dan Tian, and Wei Lü (figures 2 and 3). The leaky body includes all sickness in the spiritual, mental, emotional, and physical bodies. Shen Qi Jing blockages in these major spaces can cause many sicknesses. To purify and clear these spaces is one of the keys to fulfilling the leaky body, which is to heal and prevent all kinds of sickness.

Each Soul House also has its unique significance for the soul journey. Each one is directly related to your spiritual standing. The higher the Soul House in which your soul resides, the higher your spiritual standing in Heaven. The soul journey is to uplift your soul to reside in higher and higher Soul Houses within your body.

The Qi Channel is the most important energy channel in the body. It is the key for healing all kinds of sicknesses. The Jing Channel is the most important matter channel in the body. It is the key for rejuvenation. The Shen Channel is the most important channel for the soul, heart, and mind. It is the key for longevity and immortality.

Ming Men is divided into the Ming Men acupuncture point and Ming Men area. The Ming Men area is the area inside the lower back between the kidneys. The Ming Men acupuncture point is the point on the lower back directly behind the navel. The Ming Men acupuncture point is the core of the Ming Men area. The Ming Men area is the hub of Shen Qi Jing of the five Zang (Yin) organs, which are the liver, heart, spleen, lungs, and kidneys, and the six Fu (Yang) organs, which are the gallbladder, small intestine, stomach, large intestine, urinary bladder, and San Jiao (the pathway of Qi and bodily fluid).

The Lower Dan Tian is located in the lower abdomen, centered 1.5 cun below the navel (one cun equals approximately 0.3 cm) and 2.5 cun inside the body. It is about the size of a fist. The Lower Dan Tian is one of the foundational energy centers of the body.

The Wei Lü is the tailbone or sacrum. It is another key area in the body for foundational energy.

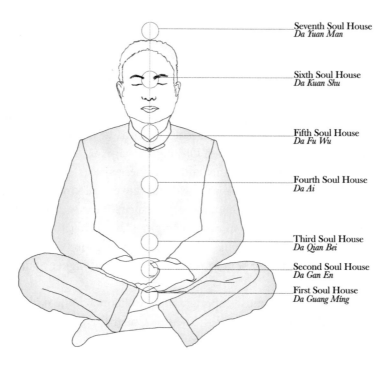

Seventh Soul House
Da Yuan Man

Sixth Soul House
Da Kuan Shu

Fifth Soul House
Da Fu Wu

Fourth Soul House
Da Ai

Third Soul House
Da Qian Bei

Second Soul House
Da Gan En

First Soul House
Da Guang Ming

Figure 2. Seven Soul Houses with associated Ten Da Tao Natures

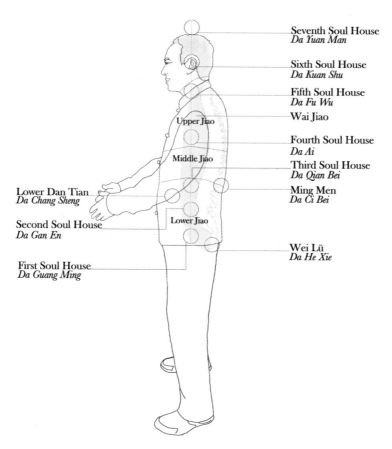

Figure 3. Seven Soul Houses, San Jiao, Wai Jiao, Ming Men point, Lower Dan Tian, and Wei Lü with associated Ten Da Tao Natures

Chapter One

Chinese	Pinyin	English Translation
嘿嘿嘿	Hei Hei Hei	*Hei Hei Hei*
嘿金丹	Hei Jin Dan	*Hei Golden Light Ball*
大光明	Da Guang Ming	*Greatest Light*
大光明金丹	Da Guang Ming Jin Dan	*Greatest Light Golden Light Ball*
我在道光中	Wo Zai Dao Guang Zhong	*I am inside Tao Source light*
道光在我中	Dao Guang Zai Wo Zhong	*Tao Source light is inside me*
通體透明	Tong Ti Tou Ming	*Whole body is bright and transparent in the Tao light*

Sacred Wisdom

嘿嘿嘿 Hei Hei Hei (First Soul House)

Hei Hei Hei is a sacred and secret Tao mantra to develop the first Soul House. The first Soul House is a fist-sized space just above the perineum, which is the region between the genitals and the anus (figure 4). At the physical level, this area is the generator of energy and life. It is the engine that produces the driving force to generate and propel energy through the remaining six Soul Houses and the entire body. At the spiritual level, the first Soul House is the generating source of energy and light. It is the beginning of the soul journey. Each human being's soul has experienced many lifetimes in order to progress on its journey. If one's soul sits in the first Soul House, the soul is still at the beginning stage of its journey. To do practice for the first Soul House is to uplift the soul's standing in Heaven and to remove significant physical and emotional blockages in life.

嘿金丹 Hei Jin Dan

"Jin" means *gold*. "Dan" means *light ball*. Ancient wisdom says, "Jin Guang Zhao Ti, Bai Bing Xiao Chu (金光照體, 百病消除)." "Jin" means *gold*. "Guang" means *light*. "Zhao" means *to shine*. "Ti" means *body*. "Bai" means *one hundred*. In this context, one hundred represents *innumerable*. "Bing" means *sickness*. "Xiao Chu" means *to remove*. "Jin Guang Zhao Ti, Bai Bing Xiao Chu" means *golden light shines in the body, all sickness is removed*.

When you chant *Hei Hei Hei, Hei Jin Dan*, you gather the Shen Qi Jing from countless planets, stars, galaxies, and universes, Tao, and Oneness to form the first Soul House Hei Jin Dan. "Hei" is a sacred mantra. "Hei Jin Dan" is a golden light ball.

In this new Tao Classic, each Soul House and several spaces, as well as various organs, has its own sacred mantra. These sacred mantras create Jin Dans (golden light balls) in their associated areas.

I would like to share with you that I have trained Tao Master Teachers and Grandmaster Teachers to transmit Tao Jin Dans to the Soul Houses, important spaces, and key organs in the body. You can receive Jin Dan transmissions from these certified Tao Masters to accelerate your physical journey and your soul journey. It would take much longer, even an entire lifetime, to form a Jin Dan solely with self-practice.

大光明 Da Guang Ming

"Da" means *greatest*. "Guang" means *light*. "Ming" means *transparent and bright*. "Da Guang Ming" means *greatest light and transparency*.

大光明金丹 Da Guang Ming Jin Dan

"Da Guang Ming Jin Dan" means *greatest light and transparency golden light ball*. When you chant *Da Guang Ming, Da Guang Ming Jin Dan*, you gather Shen Qi Jing from countless planets, stars, galaxies, and universes, Tao, and Oneness to form Da Guang Ming Jin Dan.

我在道光中 Wo Zai Dao Guang Zhong

"Wo" means *I*. "Zai" means *to be in*. "Dao Guang" means *Tao light*. "Zhong" means *within*. "Wo Zai Dao Guang Zhong" means *I am inside Tao light*.

道光在我中 Dao Guang Zai Wo Zhong

"Dao Guang Zai Wo Zhong" means *Tao light is inside me*.

通體透明 Tong Ti Tou Ming

"Tong" means *whole*. "Ti" means *body*. "Tou Ming" means *bright and transparent*. "Tong Ti Tou Ming" means *the whole body is bright and transparent in the Tao light*.

Deeper wisdom of Da Guang Ming

Da Guang Ming, greatest light, is the fourth of the very important Ten Da qualities of Tao. In fact, all forms of spiritual practice aim to develop a light body. In one sentence:

**To do spiritual practice is to transform the
physical body to a light body.**

In my book *Soul Over Matter*, four sacred phrases have been extracted from the sacred wisdom of Da Guang Ming:

Si Da Guang Ming (四大光明)
Wo Zai Dao Guang Zhong (我在道光中)
Dao Guang Zai Wo Zhong (道光在我中)
Tong Ti Tou Ming (通體透明)

The fourth of the Ten Da natures of Tao is greatest light and transparency.
I am inside Tao Source light.
Tao Source light is inside me.
The whole body is completely light and transparent.

To transform the physical body to a light body is a very slow and long process. It takes hours of practice a day for decades to make a little progress. In addition, light bodies have different frequencies and vibrations. For a Tao practitioner, there are unlimited frequencies and vibrations of light body to be attained, including Mother Earth light's frequency and vibration, Heaven light's frequency and vibration, countless planets, stars, galaxies, and universes light's frequency and vibration, and Tao Source light's frequency and vibration. Therefore, remember that the light body has unlimited layers. Tao is bigger than biggest. As Tao practitioners, we need to practice with our full Shen Qi Jing in order to reach Tao. In ancient wisdom there are sacred phrases to explain how to practice:

Yong Xin Song Nian (用心誦念)
Yong Xin Xiu Lian (用心修煉)
Xiu Xin Yang Xing (修心養性)

Yong Xin Song Nian
"Yong" means *to use*. "Xin" means *heart*. "Song Nian" means *to chant*. "Yong Xin Song Nian" means *use the heart to chant*. Using the heart yields hugely different results than not using the heart. In our book *Soul Mind Body Science System*,[6] Dr. Rulin Xiu and I have stated that Shen includes soul, heart, and mind. In quantum science, soul is information or message. Heart is the receiver. Mind is the processor. Qi is energy, which is the mover. Jing is matter, which is the transformer.

A mantra is a message or information. Heart is the receiver. Opening the heart is the key to processing the message. If one opens the heart only ten percent, then the person can receive only ten percent of the benefits. If one opens the heart fifty percent, the person can receive fifty percent of the benefits. If one opens the heart fully, the person will receive full benefits. Therefore, the sacred wisdom is that when a person chants a mantra with full concentration and greatest honor, respect, and appreciation, then the person could receive full benefits from the chanting. So, when you chant every sacred phrase in this new Tao Classic, use your heart to chant.

There is another ancient phrase, "Shi Ban Gong Bei (事半功倍)." "Shi" means *things*. "Ban" means *half*. "Gong" means *results*. "Bei" means *double* or *multiple*. "Shi Ban Gong Bei" means *half the effort can yield double or multiple results*. For example, if you chant for ten minutes absentmindedly, you would receive a little benefit. If you use the heart to chant for five minutes, you could receive twice, thrice, or even more times the benefits. This is the ancient wisdom of chanting.

---

6 Dr. and Master Zhi Gang Sha and Dr. Rulin Xiu, *Soul Mind Body Science System: Grand Unification Theory and Practice for Healing, Rejuvenation, Longevity, and Immortality*, Dallas/Toronto: BenBella Books/Heaven's Library, 2014.

Yong Xin Xiu Lian

"Xiu Lian" means *purification practice*. "Yong Xin Xiu Lian" means *use the heart to do purification practice*. What is purification? Purification is to purify our Shen Kou Yi, which are our actions, speech, and thoughts. Everything we do is through Shen Kou Yi. How do we know we are doing the right things? In one sentence:

**To serve others by teaching and serving them to have good
health and happiness and transforming blockages in any
aspect of their lives is the highest Xiu Lian practice.**

Xiu Xin Yang Xing

"Xiu" means *to purify*. "Xin" means *heart*. "Yang" means *nourish and transform*. "Xing" means *nature*. "Xiu Xin Yang Xing" means *purify the heart, nourish and transform one's nature*. Some people have the nature of love. Through Xiu Xin Yang Xing, their love can be developed deeper. Some people have the nature of compassion. Through Xiu Xin Yang Xing, their compassion can also be developed deeper. Some people can be somehow selfish. Through Xiu Xin Yang Xing, selfishness can be transformed to generosity. Some people can be mean. This can be transformed to kindness. Purifying the heart can transform one's nature.

Another ancient wisdom is Jiang Shan Yi Gai, Ben Xing Nan Yi (江山易改, 本性難移). "Jiang Shan" means *rivers and mountains*. "Yi" means *easy*. "Gai" means *to change* or *to transform*. "Ben" means *original*. "Xing" means *nature of a human being*. "Nan" means *difficult*. "Yi" means *to transform*. "Jiang Shan Yi Gai, Ben Xing Nan Yi" means *it is easier to change and transform rivers and mountains than one's nature*. To transform the nature of humanity and wan ling (all souls) is the biggest, highest transformation. Therefore, to chant and do spiritual practice, one must be patient and persistent. A Tao practitioner must have the courage, patience, persistence, and dedication to do spiritual practice. For any success in any aspect of life, persistence is a key. Finally, persistence will bring success.

Da Guang Ming is one of the highest achievements of all transformation. When you have blockages in health, relationships, finances, intelligence, or any aspect of life, your frequency and vibration are low. Da Guang Ming will increase your frequency and vibration to transform your health and build a light body. It also transforms your relationships and finances to a light body. In one sentence:

**To transform health, relationships, finances,
intelligence, and success, and to reach Ming Xin Jian
Xing, longevity, and immortality are to transform
the normal body to the greatest light body.**

I am inside Tao Source light.
Tao Source light is inside me.

This is to invoke Tao Source light, which carries Tao Shen Qi Jing that can transform our own Shen Qi Jing. Tao Source is the Ultimate Creator. Tao Source Shen Qi Jing is the purest Shen Qi Jing. When you chant, "I am inside Tao Source light. Tao Source light is inside me," you are actually invoking Tao Source to transform your Shen Qi Jing blockages. The secret wisdom is that Tao Source light transforms your Shen Qi Jing in a fast way, to serve others. To serve is to make others happier and healthier.

Tao Source transforms us. We must remember to serve more. The more we serve, the more transformation we will receive from Tao Source. Therefore, chant more.

Practice

Apply the Four Power Techniques:

1. Body Power

Two body positions can be used for the practice: sitting position (figure 4) and standing position. The sitting position is for Yin practice. The standing position is for Yang practice. Yin practice is mainly to create water energy and Yang practice is to create fire energy. Yin and Yang energy joining as one is called Yin Yang He Yi (陰陽合一).

Please also follow this principle of Yin Yang balance: Chant once in the sitting position. Then chant once in the standing position. Continue to alternate the sitting and standing positions. Or you can practice entirely in the sitting position. In this case, do your next practice entirely in the standing position.

Always alternate sitting and standing positions in your practice. This is the sacred principle of Yin Yang balance to achieve the best results.

Here is the sitting position: Sit up straight. Raise the tip of your tongue so that it gently touches the roof of your mouth. Hold your left thumb firmly with your right hand and place both hands over your first Soul House at the bottom of your abdomen.

Here is the standing position: Stand with your feet shoulder-width apart. Bend your knees slightly. Tuck in your abdomen and pelvis. Hold your arms and hands in front of you as though you were holding a ball at the level of your navel. Keep your back straight. Tuck in your chin slightly. Imagine there is a string on the top of your head pulling you gently upward.

2. Soul Power

Say *hello* to inner and outer souls:

> *Dear Tao Source,*
> *Dear Heaven and Mother Earth,*
> *Dear countless planets, stars, galaxies, and universes,*
> *Dear Bao Yuan Shou Yi Chang Shou Yong Sheng Dao Jing* (Tao Classic of Longevity and Immortality),
> *I love you, honor you, and appreciate you.*
> *Please purify and remove my Shen Qi Jing blockages for healing, rejuvenation, transformation of relationships and finances, increasing intelligence, and flourishing in every aspect of my life.*
> *Please bless my spiritual practice for longevity and immortality and develop my first Soul House and my Da Guang Ming, greatest light.*
> *Please bless me to achieve longevity and immortality.*
> *I am extremely honored and grateful.*
> *Thank you. Thank you. Thank you.*

3. Mind Power

Visualize a Tao Source Jin Dan (golden light ball) shining and rotating in your first Soul House and then radiating through your entire body (figure 4). The light is cleansing Shen Qi Jing blockages in your first Soul House and then getting stronger and brighter as the Jin Dan grows to the size of your body.

Figure 4. First Soul House Jin Dan Practice: front view and side view

4. Sound Power

Chant the first stanza (chapter one) of the *Tao Classic of Longevity and Immortality*:

> *Hei Hei Hei*
> *Hei Jin Dan*
> *Da Guang Ming*
> *Da Guang Ming Jin Dan*
> *Wo Zai Dao Guang Zhong*
> *Dao Guang Zai Wo Zhong*
> *Tong Ti Tou Ming*

Chant aloud or silently at least ten times—the more times the better. If you want to accelerate your healing and your Tao journey, you can chant for a half hour or even one to two hours for this and each Soul House. There is no time limit. To chant aloud is Yang chanting, which creates fire energy. To chant silently is Yin chanting, which creates water energy. Alternate chanting aloud and chanting silently to maximize the benefits.

Chapter Two

Chinese	Pinyin	English Translation
哼哼哼	Heng Heng Heng	*Heng Heng Heng*
哼金丹	Heng Jin Dan	*Heng Golden Light Ball*
大感恩	Da Gan En	*Greatest Gratitude*
大感恩金丹	Da Gan En Jin Dan	*Greatest Gratitude Golden Light Ball*
道生德養	Dao Sheng De Yang	*Tao Source creates and De nourishes everyone and everything*
栽培賜慧	Zai Pei Ci Hui	*Tao Source cultivates and bestows wisdom*
道恩永存	Dao En Yong Cun	*The glory of Tao Source is eternal*

Sacred wisdom

哼哼哼 Heng Heng Heng (Second Soul House)
Heng is a sacred and secret Tao mantra and esoteric wisdom to develop the second Soul House. The second Soul House is located in the lower abdomen between the first Soul House and level of the navel (figure 5).

哼金丹 Heng Jin Dan
When you chant *Heng Heng Heng, Heng Jin Dan*, you gather Shen Qi Jing from countless planets, stars, galaxies, and universes, Tao, and Oneness to form the second Soul House Jin Dan. It is called Heng Jin Dan.

13

大感恩 Da Gan En

"Da" means *greatest*. "Gan En" means *gratitude*. "Da Gan En" means *greatest gratitude*.

大感恩金丹 Da Gan En Jin Dan

"Da Gan En Jin Dan" means *greatest gratitude golden light ball*. When you chant *Da Gan En Jin Dan*, you are gathering Shen Qi Jing from Tao Source and Oneness, which are the Wu World, as well as from countless planets, stars, galaxies, and universes, which are the You World, to form Da Gan En Jin Dan.

道生德養 Dao Sheng De Yang

Dao is the Source and Ultimate Creator. "Sheng" means *create*. "De" means *virtue*. De is the Shen Kou Yi of Tao. "Shen" means *action*. "Kou" means *speech*. "Yi" means *thought*. "Yang" means *nourish*. "Dao Sheng De Yang" means *the Source and Ultimate Creator creates everyone and everything, and De nourishes them*.

栽培賜慧 Zai Pei Ci Hui

"Zai Pei" means *to cultivate*. "Ci" means *bestow*. "Hui" means *wisdom and intelligence*. "Zai Pei Ci Hui" means *the Source and Ultimate Creator cultivates and bestows wisdom and intelligence on everyone and everything*.

道恩永存 Dao En Yong Cun

Dao is the Source and Ultimate Creator. "En" means *glory*. "Yong Cun" means *forever* or *everlasting*. "Dao En Yong Cun" means *the glory of the Source and Ultimate Creator remains in our hearts and souls forever*.

Deeper wisdom of Da Gan En

In our lives there are many people who have helped us physically or spiritually. We should always show our gratitude to them. For instance, parents raise their children. Children should have gratitude for their parents. Teachers in kindergarten, elementary school, middle school, high school, and university teach students. Students should express gratitude to their teachers.

The Divine and Tao Source have bestowed appropriate wisdom and blessings to every soul's physical life and spiritual journey. We should appreciate and be grateful to them. Gratitude is one of the Tao natures that every human being and every soul should have. For our good health, happiness, harmonious relationships, and financial and business flourishing, we need to show our gratitude to the Divine, Tao Source, and anyone who has helped us on our spiritual or physical journeys.

The four-line sacred mantra of *Da Gan En* can transform our lives further and enhance every aspect of our lives:

Ba Da Gan En (八大感恩)
Dao Sheng De Yang (道生德養)
Zai Pei Ci Hui (栽培賜慧)
Dao En Yong Cun (道恩永存)

The eighth of the Ten Da Tao Natures is greatest gratitude.
Tao Source creates all things, De nourishes them.
Tao Source cultivates and bestows wisdom and intelligence.
The glory of Tao Source should remain in our hearts and souls forever.

Practice

Apply the Four Power Techniques:

1. Body Power

Sit or stand as in the practice in chapter one. Remember to alternate sitting and standing positions in your practice in order to achieve the best results.

2. Soul Power

Say *hello* to inner and outer souls:

Dear Tao Source,
Dear Heaven and Mother Earth,
Dear countless planets, stars, galaxies, and universes,

Dear Bao Yuan Shou Yi Chang Shou Yong Sheng Dao Jing (Tao Classic of
   Longevity and Immortality),
I love you, honor you, and appreciate you.
*Please purify and remove my Shen Qi Jing blockages for healing, rejuvenation,
   transformation of relationships and finances, increasing intelligence, and
   flourishing in every aspect of my life.*
*Please bless my spiritual practice of longevity and immortality and develop my
   second Soul House and my Da Gan En, greatest gratitude.*
Please bless me to achieve longevity and immortality.
I am extremely honored and grateful.
Thank you. Thank you. Thank you.

3. Mind Power

Visualize a Tao Source Jin Dan (golden light ball) shining and rotating in the
second Soul House and then radiating through your entire body (figure 5). The
light is cleansing Shen Qi Jing blockages in your second Soul House and then
getting stronger and brighter as the Jin Dan grows to the size of your body.

Figure 5. Second Soul House Jin Dan Practice: front view and side view

4. Sound Power

Chant silently or aloud the first two stanzas (chapters one and two) of the *Tao Classic of Longevity and Immortality*, and then repeat the second stanza (in **bold** below) ten or more times. Remember that the best results are achieved by alternating silent and aloud chanting for Yin Yang balance.

Hei Hei Hei
Hei Jin Dan
Da Guang Ming
Da Guang Ming Jin Dan
Wo Zai Dao Guang Zhong
Dao Guang Zai Wo Zhong
Tong Ti Tou Ming (1 time)

Heng Heng Heng
Heng Jin Dan
Da Gan En

Da Gan En Jin Dan
Dao Sheng De Yang
Zai Pei Ci Hui
Dao En Yong Cun (10 times)

Chapter Three

Chinese	Pinyin	English Translation
哄哄哄	Hong Hong Hong	*Hong Hong Hong*
哄金丹	Hong Jin Dan	*Hong Golden Light Ball*
大謙卑	Da Qian Bei	*Greatest Humility*
大謙卑金丹	Da Qian Bei Jin Dan	*Greatest Humility Golden Light Ball*
柔弱不爭	Rou Ruo Bu Zheng	*Be gentle and soft, do not compete*
持續精進	Chi Xu Jing Jin	*Improve persistently*
失謙卑	Shi Qian Bei	*Lose humility*
跌萬丈	Die Wan Zhang	*Fail tremendously in every aspect of life, just like falling into a deep cavern*

Sacred wisdom

哄哄哄 Hong Hong Hong (Third Soul House)
Hong is a sacred and secret Tao mantra to develop the third Soul House. The third Soul House is located at the level of the navel (figure 6).

哄金丹 Hong Jin Dan
"Hong Jin Dan" is a Hong Golden Light Ball. When you chant *Hong Hong Hong, Hong Jin Dan*, you will gather Shen Qi Jing in a secret way from countless planets, stars, galaxies, and universes, Tao, and Oneness to form the third Soul House Jin Dan. It is called Hong Jin Dan.

大謙卑 Da Qian Bei

"Qian Bei" means *humility*. "Da Qian Bei" means *greatest humility*.

大謙卑金丹 Da Qian Bei Jin Dan

"Da Qian Bei Jin Dan" means *greatest humility golden light ball*. When you chant *Da Qian Bei Jin Dan*, you are gathering Shen Qi Jing from Tao Source and Oneness, which are the Wu World, as well as from countless planets, stars, galaxies, and universes, which are the You World, to form Da Qian Bei Jin Dan in your third Soul House.

柔弱不爭 Rou Ruo Bu Zheng

"Rou" means *gentle*. "Ruo" means *soft*. "Bu" means *do not*. "Zheng" means *compete*. "Rou Ruo Bu Zheng" means *be gentle and soft, do not compete*. In ancient spiritual wisdom, a gentle and soft nature always overcomes a strong and harsh nature. For example, gently dripping water can wear through a stone to create a hole.

持續精進 Chi Xu Jing Jin

"Chi Xu" means *persistent*. "Jing Jin" means *improve, advance*. "Chi Xu Jing Jin" means *improve and advance persistently*.

失謙卑 Shi Qian Bei

"Shi" means *lose*. "Qian Bei" means *humility*. "Shi Qian Bei" means *lose humility*.

跌萬丈 Die Wan Zhang

"Die" means *fall*. "Wan" means *ten thousand*. In Chinese, "wan" represents *countless, innumerable*. "Zhang" is a Chinese measuring unit of length that equals 3.3 meters. "Die Wan Zhang" means *fall down very deeply*. Here it implies tremendous failures, obstacles, and loss in every aspect of life.

Deeper wisdom of Da Qian Bei

Da Qian Bei, greatest humility, is the fifth of the Ten Da Tao Natures. Humility benefits life and prevents disasters and challenges. Ego is one of the biggest blockages in every aspect of life. When you succeed financially or in business, or when you are in a powerful position, such as a government official or a world leader, congratulations for your success. Be aware that it is also a time when your ego could grow very easily. Use the greatest humility to contain your ego.

An ancient statement can help you understand why humility is important: Mo Shi Zai Ren, Cheng Shi Zai Tian (謀事在人, 成事在天). "Mo" means *planning, marketing, organizing, controlling, developing, and more.* "Shi" means *thing.* "Zai" means *depend.* "Ren" means *human being.* "Cheng" means *success.* "Tian" means *Heaven.* "Mo Shi Zai Ren, Cheng Shi Zai Tian" means *planning, marketing, and developing anything depend on a human team, but success depends on Heaven's team.*

A human team and Heaven's team are the Yin team and the Yang team. A human team is in the You World. Heaven's team is in the Wu World. When you reach success in business, in politics, or in any aspect of life, remain humble. Then Heaven's team will continue to bless you with further flourishing. If you grow ego, you will lose the blessings from Heaven's team. You could learn a serious lesson.

Why do some businesses flourish more and more? Why do some businesses get smaller and smaller and even go bankrupt? Why are some country and world leaders always honored by people, while some are not? People have not realized the sacred wisdom that humility cultivates one's good health and happiness. Modern medicine and many other healing modalities have not realized the wisdom that ego is one of the major causes of sickness.

Therefore, humility is the key for continued success in business, health, social and all kinds of relationships, and every aspect of life. To chant the few sacred phrases of Da Qian Bei is to remind you to remain humble and to avoid ego. Your life will be blessed.

Practice

Apply the Four Power Techniques:

1. Body Power

Sit or stand as in the practice in chapter one. Remember to alternate sitting and standing positions in your practice in order to achieve the best results.

2. Soul Power

Say *hello* to inner and outer souls:

> *Dear Tao Source,*
> *Dear Heaven and Mother Earth,*
> *Dear countless planets, stars, galaxies, and universes,*
> *Dear Bao Yuan Shou Yi Chang Shou Yong Sheng Dao Jing* (Tao Classic of
>   Longevity and Immortality),
> *I love you, honor you, and appreciate you.*
> *Please purify and remove my Shen Qi Jing blockages for healing, rejuvenation,*
>   *transformation of relationships and finances, increasing intelligence, and*
>   *flourishing in every aspect of my life.*
> *Please bless my spiritual practice of longevity and immortality and develop my*
>   *third Soul House and my Da Qian Bei, greatest humility.*
> *Please bless me to achieve longevity and immortality.*
> *I am extremely honored and grateful.*
> *Thank you. Thank you. Thank you.*

3. Mind Power

Visualize a Tao Source Jin Dan (golden light ball) shining and rotating in the third Soul House and then radiating its light through your entire body (figure 6). The light cleanses Shen Qi Jing blockages in your third Soul House and gets stronger and brighter as the Jin Dan grows to the size of your body.

4. Sound Power

Chant silently or aloud the first three stanzas of the *Tao Classic of Longevity and Immortality* (from chapters one through three), and then repeat the third stanza (in **bold** below) ten or more times in each practice session. Remember that the best results are always to alternate silent (Yin) and aloud (Yang) chanting.

> *Hei Hei Hei*
> *Hei Jin Dan*

Figure 6. Third Soul House Jin Dan Practice: front view and side view

Da Guang Ming
Da Guang Ming Jin Dan
Wo Zai Dao Guang Zhong
Dao Guang Zai Wo Zhong
Tong Ti Tou Ming

Heng Heng Heng
Heng Jin Dan
Da Gan En
Da Gan En Jin Dan
Dao Sheng De Yang
Zai Pei Ci Hui
Dao En Yong Cun (1 time)

Hong Hong Hong
Hong Jin Dan
Da Qian Bei
Da Qian Bei Jin Dan
Rou Ruo Bu Zheng
Chi Xu Jing Jin
Shi Qian Bei
Die Wan Zhang (10 times)

Chapter Four

Chinese	Pinyin	English Translation
啊啊啊	Ah Ah Ah	*Ah Ah Ah*
啊金丹	Ah Jin Dan	*Ah Golden Light Ball*
大愛	Da Ai	*Greatest Love*
大愛金丹	Da Ai Jin Dan	*Greatest Love Golden Light Ball*
無條件愛	Wu Tiao Jian Ai	*Unconditional love*
融化災難	Rong Hua Zai Nan	*Melts all blockages*
心清神明	Xin Qing Shen Ming	*Purifies heart and enlightens soul, heart, and mind*

Sacred wisdom

啊啊啊 Ah Ah Ah (Fourth Soul House)
Ah is a sacred and secret Tao mantra to develop the fourth Soul House in a secret way. The fourth Soul House is located at the Message Center or heart chakra behind the sternum (figure 7).

啊金丹 Ah Jin Dan
When you chant *Ah Ah Ah, Ah Jin Dan*, you gather Shen Qi Jing in a secret way from countless planets, stars, galaxies, and universes, Tao, and Oneness to form the fourth Soul House Jin Dan, which is called Ah Jin Dan.

大愛 Da Ai
"Da" means *greatest*. "Ai" means *love*. "Da Ai" means *greatest love*.

大愛金丹 Da Ai Jin Dan

"Da Ai Jin Dan" means *greatest love golden light ball*. When you chant *Da Ai Jin Dan*, you are gathering Shen Qi Jing from Tao Source and Oneness, which are the Wu World, as well as from countless planets, stars, galaxies, and universes, which are the You World, to form Da Ai Jin Dan.

無條件愛 Wu Tiao Jian Ai

"Wu" means *no*. "Tiao Jian" means *condition*. "Ai" means *love*. "Wu Tiao Jian Ai" means *unconditional love*.

融化災難 Rong Hua Zai Nan

"Rong Hua" means to *melt* or *avoid*. "Zai Nan" means *all kinds of disasters and catastrophes*. "Rong Hua Zai Nan" means *avoid all kinds of disasters and catastrophes*.

心清神明 Xin Qing Shen Ming

"Xin" means *heart*. "Qing" means *clear and transparent*. "Shen" includes *soul, heart, and mind*. "Ming" means *enlighten*. "Xin Qing Shen Ming" means *heart is clear and transparent, and then soul, heart, and mind are enlightened.*

Deeper wisdom of Da Ai

Da Ai, greatest love, is the first of the Ten Da natures of Tao. The four-line sacred mantra of Da Ai is:

Yi Shi Da Ai (一是大愛)
Wu Tiao Jian Ai (無條件愛)
Rong Hua Zai Nan (融化災難)
Xin Qing Shen Ming (心清神明)

First of the Ten Da Tao Natures is greatest love.
Unconditional love
Melts all blockages.
Heart is clear; soul, heart, and mind are enlightened.

Millions of people have studied Buddhism. Millions of people have studied Taoism. Millions of people have studied Confucianism. Buddhism, Taoism, and Confucianism are three pillars of Chinese culture and philosophy. Also, millions of people have studied Christianity, Judaism, Hinduism, Islam, and more. I honor all spiritual groups and their teachings. I have realized that the Ten Da are the essence of all of their teachings. Therefore, to study, chant, and apply the Ten Da in every aspect of life is to serve humanity and all souls.

Greatest love is the key for a human being and all souls. Greatest love is unconditional love and selfless love. Everyone needs greatest love. Everyone appreciates greatest love. Everyone's hearts and souls are touched when receiving greatest love. Why do millions of people follow Buddhas, bodhisattvas, holy saints, Taoist saints, and other saints? It is because they give their greatest love to humanity and all souls.

Why can't all human beings offer Da Ai to one another? It is because of Shen Qi Jing blockages. To chant *Da Ai* is to remove Shen Qi Jing blockages. Da Ai carries the greatest power to remove Shen Qi Jing blockages in health, relationships, finances, success, soul enlightenment, Ming Xin Jian Xing, longevity, immortality, and all life. Therefore, chant more. Practice more. Open your heart and soul. Receive unlimited blessings of Da Ai to reach Ming Xin Jian Xing, including soul enlightenment, heart enlightenment, and mind enlightenment.

Practice

Apply the Four Power Techniques:

1. Body Power

Sit or stand as in the practice in chapter one. Remember to alternate sitting and standing positions in your practice to achieve the best results.

2. Soul Power

Say *hello* to inner and outer souls:

> *Dear Tao Source,*
> *Dear Heaven and Mother Earth,*

Dear countless planets, stars, galaxies, and universes,
Dear Bao Yuan Shou Yi Chang Shou Yong Sheng Dao Jing (Tao Classic of
*   Longevity and Immortality),*
I love you, honor you, and appreciate you.
Please purify and remove my Shen Qi Jing blockages for healing, rejuvenation,
*   transformation of relationships and finances, increasing intelligence, and*
*   flourishing in every aspect of my life.*
Please bless my spiritual practice of longevity and immortality and develop my
*   fourth Soul House and my Da Ai, greatest love.*
Please bless me to achieve longevity and immortality.
I am extremely honored and grateful.
Thank you. Thank you. Thank you.

3. Mind Power

Visualize a Tao Source Jin Dan (golden light ball) shining and rotating in your fourth Soul House and then radiating its light through your entire body (figure 7). The light is cleansing Shen Qi Jing blockages in your fourth Soul House and then getting stronger and brighter as the Jin Dan grows to the size of your body.

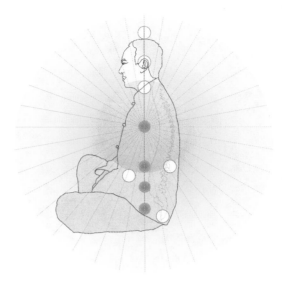

Figure 7. Fourth Soul House Jin Dan Practice: front view and side view

4. Sound Power

Chant silently or aloud stanzas one through four of the *Tao Classic of Longevity and Immortality* (from chapters one through four), and then repeat the fourth stanza (in **bold** below) ten or more times in each practice session. Remember that the best chanting practice is always to alternate silent and aloud (Yin Yang) chanting.

Hei Hei Hei
Hei Jin Dan
Da Guang Ming
Da Guang Ming Jin Dan
Wo Zai Dao Guang Zhong
Dao Guang Zai Wo Zhong
Tong Ti Tou Ming

Heng Heng Heng
Heng Jin Dan

Da Gan En
Da Gan En Jin Dan
Dao Sheng De Yang
Zai Pei Ci Hui
Dao En Yong Cun

Hong Hong Hong
Hong Jin Dan
Da Qian Bei
Da Qian Bei Jin Dan
Rou Ruo Bu Zheng
Chi Xu Jing Jin
Shi Qian Bei
Die Wan Zhang (1 time)

Ah Ah Ah
Ah Jin Dan
Da Ai
Da Ai Jin Dan
Wu Tiao Jian Ai
Rong Hua Zai Nan
Xin Qing Shen Ming (10 times)

Chapter Five

Chinese	Pinyin	English Translation
唏唏唏	Xi Xi Xi	*Xi Xi Xi*
唏金丹	Xi Jin Dan	*Xi Golden Light Ball*
大服務	Da Fu Wu	*Greatest Service*
大服務金丹	Da Fu Wu Jin Dan	*Greatest Service Golden Light Ball*
誓為公僕	Shi Wei Gong Pu	*Vow to be a servant of humanity, Mother Earth, and Heaven*
無私奉獻	Wu Si Feng Xian	*Serve selflessly*
上乘法門	Shang Cheng Fa Men	*The highest way to reach Tao Source*

Sacred wisdom

唏唏唏 Xi Xi Xi (Fifth Soul House)
Xi is a sacred and secret Tao mantra to develop the fifth Soul House. The fifth Soul House is located in the throat (figure 8).

唏金丹 Xi Jin Dan
When you chant *Xi Xi Xi, Xi Jin Dan,* you gather Shen Qi Jing in a secret way from countless planets, stars, galaxies, and universes, Tao, and Oneness to form the fifth Soul House Jin Dan, which is called Xi Jin Dan.

大服務 Da Fu Wu
"Da" means *greatest*. "Fu Wu" means *service*. "Da Fu Wu" means *greatest service*.

大服務金丹 Da Fu Wu Jin Dan

"Da Fu Wu Jin Dan" means *greatest service golden light ball*. When you chant *Da Fu Wu Jin Dan*, you are gathering Shen Qi Jing from Tao Source and Oneness, which are the Wu World, as well as from countless planets, stars, galaxies, and universes, which are the You World, to form Da Fu Wu Jin Dan.

誓為公僕 Shi Wei Gong Pu

"Shi" means *vow*. "Wei" means *to be*. "Gong Pu" means *selfless servant*. "Shi Wei Gong Pu" means *vow to be a selfless servant of humanity, Mother Earth, and Heaven*.

無私奉獻 Wu Si Feng Xian

"Wu Si" means *selfless*. "Feng Xian" means *to dedicate* or *devote*. "Wu Si Feng Xian" means *serve humanity and all souls selflessly*.

上乘法門 Shang Cheng Fa Men

"Shang Cheng" means *highest*. "Fa" means *sacred method*. "Men" means *gate*. "Shang Cheng Fa Men" means *the highest way and method to reach Tao Source*.

Deeper wisdom of Da Fu Wu

Da Fu Wu, greatest service, is the ninth of the Ten Da natures of Tao. I always share with my students and humanity that the purpose of life is to serve. To serve is to make others happier and healthier, as well as successful in every aspect of their lives and our lives.

The four-line sacred mantra of Da Fu Wu is:

Jiu Da Fu Wu (九大服務)
Shi Wei Gong Pu (誓為公僕)
Wu Si Feng Xian (無私奉獻)
Shang Cheng Fa Men (上乘法門)

The ninth of the Ten Da Tao Natures is greatest service.
Vow to be a servant of humanity and all souls.
Selflessly offer service.
The highest way to reach Tao Source.

I would like to share a personal story from my spiritual journey. Many years ago, when I was in Taiwan, Shakyamuni Buddha, the founder of Buddhism, appeared to me one day during my meditation. He is one of my spiritual fathers forever. I asked him, "Shi Jia Mo Ni Fo (釋迦牟尼佛), you taught 84,000 methods of practice. Which is the number one method?" He responded, "What do you think?" I said, "I feel the number one method for the Xiu Lian journey should be service—to serve others, to make others happier and healthier." He smiled at me and said, "I couldn't agree with you more." Service has different layers. Serve a little, serve more, or serve unconditionally. To serve unconditionally is the highest way to progress on your spiritual journey.

Practice

Apply the Four Power Techniques:

1. Body Power

Sit or stand as in the practice in chapter one. Remember to alternate sitting and standing positions in your practice in order to achieve optimum results.

2. Soul Power

Say *hello* to inner and outer souls:

> *Dear Tao Source,*
> *Dear Heaven and Mother Earth,*
> *Dear countless planets, stars, galaxies, and universes,*
> *Dear Bao Yuan Shou Yi Chang Shou Yong Sheng Dao Jing* (Tao Classic of Longevity and Immortality),
> *I love you, honor you, and appreciate you.*
> *Please purify and remove my Shen Qi Jing blockages for healing, rejuvenation, transformation of relationships and finances, increasing intelligence, and flourishing in every aspect of my life.*

Please bless my spiritual practice of longevity and immortality and develop my
      fifth Soul House and my Da Fu Wu, greatest service.
Please bless me to achieve longevity and immortality.
I am extremely honored and grateful.
Thank you. Thank you. Thank you.

3. Mind Power

Visualize a Tao Source Jin Dan (golden light ball) shining and rotating in the fifth Soul House and then radiating light through your entire body (figure 8). The light is cleansing Shen Qi Jing blockages in your fifth Soul House and then getting stronger and brighter as the Jin Dan grows to the size of your body.

4. Sound Power

Chant silently or aloud stanzas one through five of the *Tao Classic of Longevity and Immortality* (from chapters one through five), and then repeat the fifth stanza (in **bold** below) ten or more times in each practice session. Remember that the best results are obtained by alternating silent and aloud (Yin Yang) chanting.

Hei Hei Hei
Hei Jin Dan
Da Guang Ming
Da Guang Ming Jin Dan
Wo Zai Dao Guang Zhong
Dao Guang Zai Wo Zhong
Tong Ti Tou Ming

Heng Heng Heng
Heng Jin Dan
Da Gan En
Da Gan En Jin Dan
Dao Sheng De Yang
Zai Pei Ci Hui

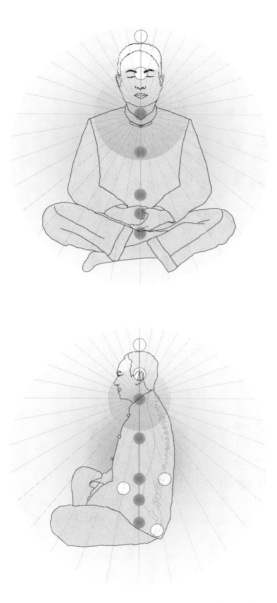

Figure 8. Fifth Soul House Jin Dan Practice: front view and side view

Dao En Yong Cun

Hong Hong Hong
Hong Jin Dan
Da Qian Bei
Da Qian Bei Jin Dan
Rou Ruo Bu Zheng
Chi Xu Jing Jin
Shi Qian Bei
Die Wan Zhang

Ah Ah Ah
Ah Jin Dan
Da Ai
Da Ai Jin Dan
Wu Tiao Jian Ai
Rong Hua Zai Nan
Xin Qing Shen Ming (1 time)

Xi Xi Xi
Xi Jin Dan
Da Fu Wu
Da Fu Wu Jin Dan
Shi Wei Gong Pu
Wu Si Feng Xian
Shang Cheng Fa Men (10 times)

Chapter Six

Chinese	Pinyin	English Translation
噎噎噎	Yi Yi Yi	*Yi Yi Yi*
噎金丹	Yi Jin Dan	*Yi Golden Light Ball*
大寬恕	Da Kuan Shu	*Greatest Forgiveness*
大寬恕金丹	Da Kuan Shu Jin Dan	*Greatest Forgiveness Golden Light Ball*
我原諒你	Wo Yuan Liang Ni	*I forgive you*
你原諒我	Ni Yuan Liang Wo	*You forgive me*
相愛平安和諧	Xiang Ai Ping An He Xie	*Love, peace, and harmony*

Sacred wisdom

噎噎噎 Yi Yi Yi (Sixth Soul House)
Yi is a sacred and secret Tao mantra to develop the sixth Soul House. The sixth Soul House is located in the brain (figure 9).

噎金丹 Yi Jin Dan
When you chant *Yi Yi Yi, Yi Jin Dan*, you gather Shen Qi Jing in a secret way from countless planets, stars, galaxies, and universes, Tao, and Oneness to form the sixth Soul House Jin Dan, which is called Yi Jin Dan.

大寬恕 Da Kuan Shu
"Da" means *greatest*. "Kuan Shu" means *forgiveness*. "Da Kuan Shu" means *greatest forgiveness*.

大寬恕金丹 Da Kuan Shu Jin Dan

"Da Kuan Shu Jin Dan" means *greatest forgiveness golden light ball*. When you chant *Da Kuan Shu Jin Dan*, you are gathering Shen Qi Jing from Tao Source and Oneness, which are the Wu World, as well as from countless planets, stars, galaxies, and universes, which are the You World, to form Da Kuan Shu Jin Dan.

我原諒你 Wo Yuan Liang Ni

"Wo" means *I*. "Yuan Liang" means *to forgive*. "Ni" means *you*. "Wo Yuan Liang Ni" means *I forgive you*.

你原諒我 Ni Yuan Liang Wo

"Ni Yuan Liang Wo" means *you forgive me*.

相愛平安和諧 Xiang Ai Ping An He Xie

"Xiang Ai" means *love each other*. "Ping An" means *peace*. "He Xie" means *harmony*. "Xiang Ai Ping An He Xie" means *love, peace, and harmony*.

Deeper wisdom of Da Kuan Shu

Da Kuan Shu, greatest forgiveness, is the second of the Ten Da natures of Tao. The four-line sacred mantra of Da Kuan Shu is:

> Er Da Kuan Shu (二大寬恕)
> Wo Yuan Liang Ni (我原諒你)
> Ni Yuan Liang Wo (你原諒我)
> Xiang Ai Ping An He Xie (相愛平安和諧)

> *Second of the Ten Da Tao Natures is greatest forgiveness.*
> *I forgive you.*
> *You forgive me.*
> *Love, peace, and harmony.*

Da Kuan Shu is great wisdom and practice to reach Ming Xin Jian Xing. Everyone has experienced some challenges, including anger, depression, anxiety, worry, sadness, fear, and all kinds of other emotions and issues. We can have challenges between husband and wife, and with relatives, colleagues, countries, and religions. We can have physical, emotional, mental, and spiritual challenges. One of the most important and most effective ways to overcome all kinds of challenges is through forgiveness.

I forgive you.
You forgive me.
Bring love, peace, and harmony.

This is a vital practice to remove anger and all other unbalanced emotions to bring love, peace, and harmony to your life. Why is a person not able to forgive others? The reason is Shen Qi Jing blockages. There is great wisdom that humanity should know. Suppose for example that colleague A treats colleague B inappropriately. B could react very angrily to A. However, if you have deep spiritual wisdom or the ability to see past lifetimes, you would understand that B had treated A in the same inappropriate way in one or more past lifetimes. With this knowledge, which is called the Law of Karma, you would react differently.

We can expand this wisdom to relatives, countries, and religions. The same principle applies. The Law of Karma is the root principle for all life. To purify and remove Shen Qi Jing blockages of all physical, emotional, mental, and spiritual challenges, the key is to forgive. I understand that it can be hard to forgive someone who hurt or harmed us. However, to truly forgive each other is the key to self-clear our own negative karma and to purify all Shen Qi Jing blockages with all kinds of emotions. It is easy to say. It is difficult to do. It is why one of the highest qualities for a true spiritual being is the ability to offer unconditional forgiveness. It is one of the highest spiritual principles and practices as well.

How can you forgive? Chant the four sacred phrases with your heart and soul. It will help you to:

- forgive more and more
- transform all kinds of emotions
- heal all kinds of sicknesses
- transform all kinds of relationships and finances
- reach Ming Xin Jian Xing.

Practice

Apply the Four Power Techniques:

1. Body Power

Sit or stand as in the practice in chapter one. Remember to alternate sitting and standing positions in your practice in order to achieve the best results.

2. Soul Power

Say *hello* to inner and outer souls:

> *Dear Tao Source,*
> *Dear Heaven and Mother Earth,*
> *Dear countless planets, stars, galaxies, and universes,*
> *Dear Bao Yuan Shou Yi Chang Shou Yong Sheng Dao Jing* (Tao Classic of
>   Longevity and Immortality),
> *I love you, honor you, and appreciate you.*
> *Please purify and remove my Shen Qi Jing blockages for healing, rejuvenation,*
>   *transformation of relationships and finances, increasing intelligence, and*
>   *flourishing in every aspect of my life.*
> *Please bless my spiritual practice of longevity and immortality and develop my*
>   *sixth Soul House and my Da Kuan Shu, greatest forgiveness.*
> *Please bless me to achieve longevity and immortality.*
> *I am extremely honored and grateful.*
> *Thank you. Thank you. Thank you.*

3. Mind Power

Visualize a Tao Source Jin Dan (golden light ball) shining and rotating in the sixth Soul House and then radiating light through the entire body (figure 9). The light is cleansing Shen Qi Jing blockages in your sixth Soul House and then getting stronger and brighter as the Jin Dan grows to the size of your body.

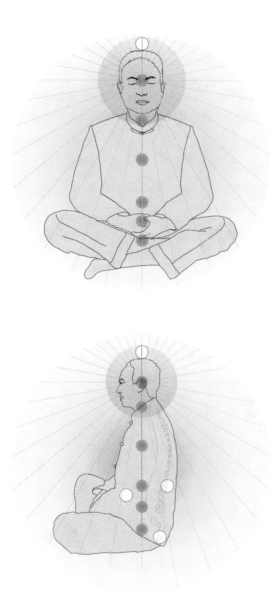

Figure 9. Sixth Soul House Jin Dan Practice: front view and side view

4. Sound Power

Chant silently or aloud stanzas one through six of the *Tao Classic of Longevity and Immortality* (from chapters one through six), and then repeat the sixth stanza (in **bold** below) ten or more times in each practice session. Remember to alternate silent and aloud (Yin Yang) chanting for best results.

> *Hei Hei Hei*
> *Hei Jin Dan*
> *Da Guang Ming*
> *Da Guang Ming Jin Dan*
> *Wo Zai Dao Guang Zhong*
> *Dao Guang Zai Wo Zhong*
> *Tong Ti Tou Ming*
>
> *Heng Heng Heng*
> *Heng Jin Dan*
> *Da Gan En*
> *Da Gan En Jin Dan*
> *Dao Sheng De Yang*
> *Zai Pei Ci Hui*
> *Dao En Yong Cun*
>
> *Hong Hong Hong*
> *Hong Jin Dan*
> *Da Qian Bei*
> *Da Qian Bei Jin Dan*
> *Rou Ruo Bu Zheng*
> *Chi Xu Jing Jin*
> *Shi Qian Bei*
> *Die Wan Zhang*
>
> *Ah Ah Ah*
> *Ah Jin Dan*
> *Da Ai*
> *Da Ai Jin Dan*
> *Wu Tiao Jian Ai*

Rong Hua Zai Nan
Xin Qing Shen Ming

Xi Xi Xi
Xi Jin Dan
Da Fu Wu
Da Fu Wu Jin Dan
Shi Wei Gong Pu
Wu Si Feng Xian
Shang Cheng Fa Men (1 time)

Yi Yi Yi
Yi Jin Dan
Da Kuan Shu
Da Kuan Shu Jin Dan
Wo Yuan Liang Ni
Ni Yuan Liang Wo
Xiang Ai Ping An He Xie (10 times)

Chapter Seven

Chinese	Pinyin	English Translation
嗡嗡嗡	Weng Weng Weng	*Weng Weng Weng*
嗡金丹	Weng Jin Dan	*Weng Golden Light Ball*
大圓滿	Da Yuan Man	*Greatest Enlightenment*
大圓滿金丹	Da Yuan Man Jin Dan	*Greatest Enlightenment Golden Light Ball*
靈心腦身圓滿	Ling Xin Nao Shen Yuan Man	*Soul, heart, mind, and body enlightenment*
人地天道神仙梯	Ren Di Tian Dao Shen Xian Ti	*Human saint, Mother Earth saint, Heaven saint, and Tao Source saint are enlightenment stairs*
服務修煉才可攀	Fu Wu Xiu Lian Cai Ke Pan	*Only through service and purification can one climb the stairs and reach all layers of enlightenment*

Sacred wisdom

嗡嗡嗡 Weng Weng Weng (Seventh Soul House)
Weng is a sacred and secret Tao mantra to develop the seventh Soul House. The seventh Soul House is located at the crown chakra on top of the head (figure 10).

嗡金丹 Weng Jin Dan
When you chant *Weng Weng Weng, Weng Jin Dan*, you gather Shen Qi Jing in a secret way from countless planets, stars, galaxies, and universes, Tao, and

Oneness to form the seventh Soul House Jin Dan, which is called Weng Jin Dan.

大圓滿 Da Yuan Man

"Da" means *greatest*. "Yuan Man" means *enlightenment*. "Da Yuan Man" means *greatest enlightenment*.

大圓滿金丹 Da Yuan Man Jin Dan

"Da Yuan Man Jin Dan" means *greatest enlightenment golden light ball*. When you chant *Da Yuan Man Jin Dan*, you are gathering Shen Qi Jing from Tao Source and Oneness, which are the Wu World, as well as from countless planets, stars, galaxies, and universes, which are the You World, to form Da Yuan Man Jin Dan.

靈心腦身圓滿 Ling Xin Nao Shen Yuan Man

"Ling" means *soul*. "Xin" means *heart*. "Nao" means *mind*. "Shen" means *body*. "Yuan Man" means *enlightenment*. "Ling Xin Nao Shen Yuan Man" means *soul, heart, mind, and body enlightenment*.

人地天道神仙梯 Ren Di Tian Dao Shen Xian Ti

"Ren" means *human being*. "Di" means *Mother Earth*. "Tian" means *Heaven*. Dao is the Source and Ultimate Creator. "Shen Xian" means *saints*. "Ti" means *stairs* or *levels*. "Ren Di Tian Dao Shen Xian Ti" means *the different levels of saints—Human saint, Mother Earth saint, Heaven saint, and Tao Source saint—are the stairs to greatest enlightenment*.

服務修煉才可攀 Fu Wu Xiu Lian Cai Ke Pan

"Fu Wu" means *service*. "Xiu Lian" means *purification practice*. "Cai" means *only*. "Ke" means *able to*. "Pan" means *to climb*. "Fu Wu Xiu Lian Cai Ke Pan" means *only through service and purification can one climb all of Heaven's enlightenment stairs step by step, from Human saint to Mother Earth saint, to Heaven saint, and finally to Tao Source saint*.

Deeper wisdom of Da Yuan Man

Da Yuan Man, greatest enlightenment, is the tenth of the Ten Da natures of Tao. It is the last Tao nature required to reach Tao. The four-line sacred mantra of Da Yuan Man is:

Shi Da Yuan Man (十大圓滿)
Ling Xin Nao Shen Yuan Man (靈心腦身圓滿)
Ren Di Tian Dao Shen Xian Ti (人地天道神仙梯)
Fu Wu Xiu Lian Cai Ke Pan (服務修煉才可攀)

The tenth Da Tao Nature is greatest enlightenment.
Soul, heart, mind, and body enlightenment
Levels of Ren Xian, Di Xian, Tian Xian, and Tao Xian
Only through service can greatest enlightenment be achieved.

A human being is made of Shen Qi Jing: soul, heart, mind, and body. To reach greatest enlightenment is to enlighten soul, heart, mind, and body.

In my book *Soul Mind Body Science System*, I shared four sacred phrases to express the sacred wisdom and truth of the relationships among soul, heart, mind, and body:

Ling Dao Xin Dao (靈到心到)
Xin Dao Yi Dao (心到意到)
Yi Dao Qi Dao (意到氣到)
Qi Dao Xue Dao (氣到血到)

Ling Dao Xin Dao

"Ling" means *soul* or *spirit*. In quantum science, soul is *information* or *message*. "Dao" means *arrive*. "Xin" means *heart*. "Ling Dao Xin Dao" means *the soul message arrives at the heart*. The heart is the receiver of the message.

Xin Dao Yi Dao

"Yi" means *consciousness*. It can be expressed as *thought*. "Xin Dao Yi Dao" means *the heart passes the message to the brain*. The brain is the processor of the message.

Yi Dao Qi Dao
"Yi Dao Qi Dao" means *the consciousness passes the message to the energy (Qi).* Energy is the mover.

Qi Dao Xue Dao
"Xue" means *blood*, which represents matter. "Qi Dao Xue Dao" means *the energy passes the message to the matter.* Matter is the transformer.

Ling Dao Xin Dao, Xin Dao Yi Dao, Yi Dao Qi Dao, and Qi Dao Xue Dao explain the relationships among Shen Qi Jing. This is the sacred wisdom and practice that Dr. Rulin Xiu and I released to humanity and wan ling (all souls) in the *Soul Mind Body Science System* book. Enlightenment follows the same process. These relationships can be expressed as:

- Ling (soul, spirit, information) – content of message
- Xin (heart, core of life) – receiver of message
- Nao (mind, consciousness) – processor of message
- Qi (energy) – mover of message
- Jing (matter) – transformer of message.

Jing is the transformer. From the soul, the message or information goes to the heart, then to the mind, next to the energy, and finally to the matter. After the matter transforms the message, the transformed message will be given as feedback to the soul. The soul will then give a new message. This is the sacred process in the whole body. The sacred enlightenment process is for the soul to reach enlightenment first, next the heart, then the mind, and finally the body, which includes energy and matter.

To reach Ming Xin Jian Xing is to reach the saints' level. To reach the saints' level is to have a saint's soul, heart, mind, and body. Saints have the following layers or levels:

- Ren Xian (Human saint)
- Di Xian (Mother Earth saint)
- Tian Xian (Heaven saint)
- Dao Xian (Tao Source saint).

When one's soul, heart, mind, and body have completely melded with Tao as one, immortality has been reached. At the Ren Xian, Di Xian, and Tian Xian levels, a saint's Shen Qi Jing melds with Tao more and more, until finally one becomes a Dao Xian. Then this one has completely melded with Tao, which is immortality.

Practice

Apply the Four Power Techniques:

1. Body Power

Sit or stand as in the practice in chapter one. Alternate sitting and standing positions in your practice for optimum results.

2. Soul Power

Say *hello* to inner and outer souls:

> *Dear Tao Source,*
> *Dear Heaven and Mother Earth,*
> *Dear countless planets, stars, galaxies, and universes,*
> *Dear Bao Yuan Shou Yi Chang Shou Yong Sheng Dao Jing* (Tao Classic of Longevity and Immortality),
> *I love you, honor you, and appreciate you.*
> *Please purify and remove my Shen Qi Jing blockages for healing, rejuvenation, transformation of relationships and finances, increasing intelligence, and flourishing in every aspect of my life.*
> *Please bless my spiritual practice of longevity and immortality and develop my seventh Soul House and my Da Yuan Man, greatest enlightenment.*
> *Please bless me to achieve longevity and immortality.*
> *I am extremely honored and grateful.*
> *Thank you. Thank you. Thank you.*

3. Mind Power

Visualize a Tao Source Jin Dan (golden light ball) shining and rotating in the seventh Soul House and then radiating through the entire body (figure 10). The light is cleansing Shen Qi Jing blockages in your seventh Soul House and then getting stronger and brighter as the Jin Dan grows to the size of your body.

4. Sound Power

Chant silently or aloud the first seven stanzas of the *Tao Classic of Longevity and Immortality* (from chapters one through seven), and then repeat the seventh stanza (in **bold** below) ten or more times in each practice session. Remember that it is best to alternate silent and aloud (Yin Yang) chanting.

Hei Hei Hei
Hei Jin Dan
Da Guang Ming
Da Guang Ming Jin Dan
Wo Zai Dao Guang Zhong
Dao Guang Zai Wo Zhong
Tong Ti Tou Ming

Heng Heng Heng
Heng Jin Dan
Da Gan En
Da Gan En Jin Dan
Dao Sheng De Yang
Zai Pei Ci Hui
Dao En Yong Cun

Hong Hong Hong
Hong Jin Dan
Da Qian Bei
Da Qian Bei Jin Dan
Rou Ruo Bu Zheng
Chi Xu Jing Jin

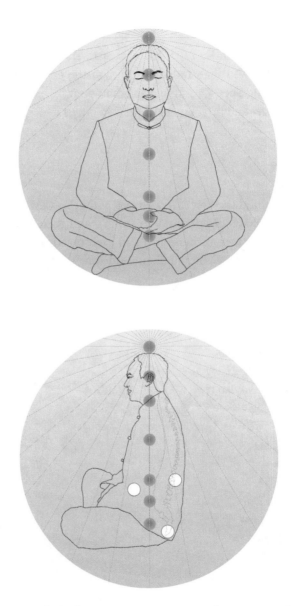

Figure 10. Seventh Soul House Jin Dan Practice: front view and side view

Shi Qian Bei
Die Wan Zhang

Ah Ah Ah
Ah Jin Dan
Da Ai
Da Ai Jin Dan
Wu Tiao Jian Ai
Rong Hua Zai Nan
Xin Qing Shen Ming

Xi Xi Xi
Xi Jin Dan
Da Fu Wu
Da Fu Wu Jin Dan
Shi Wei Gong Pu
Wu Si Feng Xian
Shang Cheng Fa Men

Yi Yi Yi
Yi Jin Dan
Da Kuan Shu
Da Kuan Shu Jin Dan
Wo Yuan Liang Ni
Ni Yuan Liang Wo
Xiang Ai Ping An He Xie (1 time)

Weng Weng Weng
Weng Jin Dan
Da Yuan Man
Da Yuan Man Jin Dan
Ling Xin Nao Shen Yuan Man
Ren Di Tian Dao Shen Xian Ti
Fu Wu Xiu Lian Cai Ke Pan (10 times)

Chapter Eight

Chinese	Pinyin	English Translation
呦呦呦	You You You	*You You You*
呦金丹	You Jin Dan	*You Golden Light Ball*
大慈悲	Da Ci Bei	*Greatest Compassion*
大慈悲金丹	Da Ci Bei Jin Dan	*Greatest Compassion Golden Light Ball*
願力增強	Yuan Li Zeng Qiang	*Increase willpower*
服務眾生	Fu Wu Zhong Sheng	*Serve humanity*
功德無量	Gong De Wu Liang	*Virtue will be immeasurable*

Sacred wisdom

呦呦呦 You You You (Ming Men)
You is a sacred and secret Tao mantra to develop the Ming Men point and area, as well as to clear the Wai Jiao (figure 11).

呦金丹 You Jin Dan
When you chant *You You You, You Jin Dan*, you gather Shen Qi Jing in a secret way from countless planets, stars, galaxies, and universes, Tao, and Oneness to clear the Wai Jiao and form a Ming Men Jin Dan, which is called You Jin Dan.

大慈悲 Da Ci Bei
"Da" means *greatest*. "Ci Bei" means *compassion*. "Da Ci Bei" means *greatest compassion*.

大慈悲金丹 Da Ci Bei Jin Dan

"Da Ci Bei Jin Dan" means *greatest compassion golden light ball*. When you chant *Da Ci Bei Jin Dan*, you are gathering Shen Qi Jing from Tao Source and Oneness, which are the Wu World, as well as from countless planets, stars, galaxies, and universes, which are the You World, to form Da Ci Bei Jin Dan.

願力增強 Yuan Li Zeng Qiang

"Yuan Li" means *willpower*. "Zeng Qiang" means *to increase and strengthen*. "Yuan Li Zeng Qiang" means *increase and strengthen willpower*.

服務眾生 Fu Wu Zhong Sheng

"Fu Wu" means *to serve*. "Zhong Sheng" means *humanity and all souls*. "Fu Wu Zhong Sheng" means *serve humanity and all souls*.

功德無量 Gong De Wu Liang

"Gong De" means *virtue*. "Wu Liang" means *immeasurable*. "Gong De Wu Liang" means *immeasurable virtue*.

Deeper wisdom of Da Ci Bei

Da Ci Bei, greatest compassion, is the third of the Ten Da natures of Tao. The sacred four-line Tao Source mantra of Da Ci Bei is:

San Da Ci Bei (三大慈悲)
Yuan Li Zeng Qiang (願力增強)
Fu Wu Zhong Sheng (服務眾生)
Gong De Wu Liang (功德無量)

The third of the Ten Da Tao Natures is greatest compassion.
Increase and strengthen willpower.
Serve humanity and all souls.
Gain immeasurable virtue.

Compassion is a great quality for a spiritual being. Guan Yin, the Bodhisattva of Compassion, is a great example and spiritual mother for millions of people. She made most powerful vows (in the Lotus Sutra) to save sentient beings from any life-threatening situations, such as fire, water, war, and sickness. When you call her name, her soul will instantly come to save you and bless you unconditionally.

One of many stories and legends about Guan Yin involves a fisherman who was fishing in the ocean. When a severe storm suddenly began, his boat capsized. The fisherman thought he was going to drown. He called Guan Yin, shouting "Guan Yin Jiu Ming (觀音救命)." "Jiu" means *save*. "Ming" means *life*. A huge wave swamped him and his boat, and he lost consciousness. When he awoke, he was lying safely on the shore. This is how Guan Yin saves people's lives. There are also many stories of people who received miraculous healing from incurable diseases when they chanted *Weng Ma Ni Ba Ma Hong*, which is Guan Yin's six-word mantra of enlightenment. That is why Guan Yin has been deeply honored by millions of people.

There are five sacred phrases to honor Guan Yin. They are:

Qian Shou Qian Yan (千手千眼)
Da Ci Da Bei (大慈大悲)
Jiu Ku Jiu Nan (救苦救難)
Guang Da Yuan Man (廣大圓滿)
Guan Shi Yin Pu Sa (觀世音菩薩)

One thousand spiritual hands, one thousand spiritual eyes
Greatest compassion
Save me and overcome all my challenges and sufferings
Greatest enlightenment
Guan Yin bodhisattva

"Qian" means *thousand*. In Chinese, a thousand represents *innumerable*. "Shou" means *hands*. "Yan" means *eyes*. "Qian Shou Qian Yan" can mean *a thousand hands and a thousand eyes* or *uncountable hands and uncountable eyes*. These hands and eyes of Guan Yin are spiritual hands and eyes. In her hands, she holds Fa Qi (法器). "Fa" means *spiritual*. "Qi" means *instruments*. Fa Qi can include swords, towers, balls, and many more. A Fa Qi carries high-dimension spiritual power to remove Shen Qi Jing blockages of a human being. Therefore, Guan Yin has been given thousands of spiritual hands and eyes to remove Shen Qi Jing blockages for health, relationships, finances, and every aspect of life. She has created countless miracle healings for all life.

"Da" means *greatest*. "Ci Bei" means *compassion*. "Da Ci Da Bei" means *greatest compassion*. Guan Yin is honored as the Bodhisattva of Compassion. Guan Yin's new name in this Soul Light Era is Ling Hui Sheng Shi. The Soul Light Era is a new universal era. It began on August 8, 2003 and will last at least fifteen thousand years. "Ling" means *soul*. "Hui" means *intelligence*. "Sheng" means *saint*. "Shi" means *servant*. "Ling Hui Sheng Shi" means *soul intelligence saint servant*. She is a bodhisattva, but she is humbly called a servant.

In my teaching, every Buddha, bodhisattva, holy saint, and healing angel, our beloved Divine, Tao, and Source are all unconditional universal servants. The purpose of life is to serve. To serve is to make others happier and healthier. Da Ci Da Bei is Guan Yin's nature. She is the great example and full embodiment of greatest compassion. "Jiu" means *save*. "Ku" literally means *bitterness, sufferings*. It also represents challenges. "Nan" means *disasters and suffering*. "Ku Nan" means *all life challenges and sufferings*. Jiu Ku Jiu Nan is to save people from all kinds of life challenges and sufferings. Guan Yin is one of the greatest beings of compassion. She serves us unconditionally. Therefore, billions of people have deeply honored and respected her. "Guang Da" means *vast*. "Yuan Man" means *enlightenment* or *perfection*. Guang Da Yuan Man is to honor Guan Yin's greatest enlightenment.

Therefore, to honor Guan Yin by chanting the five sacred phrases is to transform you to be more compassionate like Guan Yin. When you chant these five phrases, the Shen Qi Jing of Guan Yin will bless you to remove Shen Qi Jing blockages of compassion and to be more compassionate.

Practice

Apply the Four Power Techniques:

1. Body Power

Sit or stand as in the practice in chapter one. Remember to alternate sitting and standing positions in your practice in order to achieve the best results.

2. Soul Power

Say *hello* to inner and outer souls:

> *Dear Tao Source,*
> *Dear Heaven and Mother Earth,*
> *Dear countless planets, stars, galaxies, and universes,*
> *Dear Bao Yuan Shou Yi Chang Shou Yong Sheng Dao Jing* (Tao Classic of
> Longevity and Immortality),
> *I love you, honor you, and appreciate you.*
> *Please purify and remove my Shen Qi Jing blockages for healing, rejuvenation,*
> *transformation of relationships and finances, increasing intelligence, and*
> *flourishing in every aspect of my life.*
> *Please bless my spiritual practice of longevity and immortality and develop my*
> *Ming Men area and my Da Ci Bei, greatest compassion.*
> *Please bless me to achieve longevity and immortality.*
> *I am extremely honored and grateful.*
> *Thank you. Thank you. Thank you.*

3. Mind Power

Visualize a Tao Source Jin Dan (golden light ball) shining and rotating in your Ming Men area and then radiating through the entire body (figure 11). The light is cleansing Shen Qi Jing blockages in your Ming Men area and then getting stronger and brighter as the Jin Dan grows to the size of your body.

4. Sound Power

Chant silently or aloud the first eight stanzas of the *Tao Classic of Longevity and Immortality* (from chapters one through eight), and then repeat the eighth stanza (in **bold** below) ten or more times in each practice session. Remember to alternate silent and aloud (Yin Yang) chanting for best results.

> *Hei Hei Hei*
> *Hei Jin Dan*

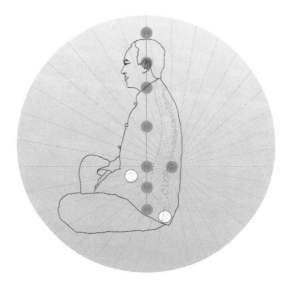

Figure 11. Ming Men Jin Dan Practice

Da Guang Ming
Da Guang Ming Jin Dan
Wo Zai Dao Guang Zhong
Dao Guang Zai Wo Zhong
Tong Ti Tou Ming

Heng Heng Heng
Heng Jin Dan
Da Gan En
Da Gan En Jin Dan
Dao Sheng De Yang
Zai Pei Ci Hui
Dao En Yong Cun

Hong Hong Hong
Hong Jin Dan
Da Qian Bei
Da Qian Bei Jin Dan
Rou Ruo Bu Zheng
Chi Xu Jing Jin

Shi Qian Bei
Die Wan Zhang

Ah Ah Ah
Ah Jin Dan
Da Ai
Da Ai Jin Dan
Wu Tiao Jian Ai
Rong Hua Zai Nan
Xin Qing Shen Ming

Xi Xi Xi
Xi Jin Dan
Da Fu Wu
Da Fu Wu Jin Dan
Shi Wei Gong Pu
Wu Si Feng Xian
Shang Cheng Fa Men

Yi Yi Yi
Yi Jin Dan
Da Kuan Shu
Da Kuan Shu Jin Dan
Wo Yuan Liang Ni
Ni Yuan Liang Wo
Xiang Ai Ping An He Xie

Weng Weng Weng
Weng Jin Dan
Da Yuan Man
Da Yuan Man Jin Dan
Ling Xin Nao Shen Yuan Man
Ren Di Tian Dao Shen Xian Ti
Fu Wu Xiu Lian Cai Ke Pan (1 time)

You You You
You Jin Dan

Da Ci Bei
Da Ci Bei Jin Dan
Yuan Li Zeng Qiang
Fu Wu Zhong Sheng
Gong De Wu Liang (10 times)

Chapter Nine

Chinese	Pinyin	English Translation
哈哈哈	Ha Ha Ha	*Ha Ha Ha*
哈金丹	Ha Jin Dan	*Ha Golden Light Ball*
大昌盛	Da Chang Sheng	*Greatest Flourishing*
大昌盛金丹	Da Chang Sheng Jin Dan	*Greatest Flourishing Golden Light Ball*
道賜盈福	Dao Ci Ying Fu	*Tao Source bestows huge fortune, including Tao wisdom, health, longevity, happiness, good relationships, prosperity, luck, success, and more to all aspects of life*
行善積德	Xing Shan Ji De	*Be kind to accumulate virtue*
道業昌盛	Dao Ye Chang Sheng	*Tao career flourishes*

Sacred wisdom

哈哈哈 Ha Ha Ha (Lower Dan Tian)
Ha is a sacred and secret Tao mantra to develop the Lower Dan Tian area. It is a crucial area in Tao teaching and practice (figure 12).

哈金丹 Ha Jin Dan
When you chant *Ha Ha Ha, Ha Jin Dan*, you gather Shen Qi Jing in a secret way from countless planets, stars, galaxies, and universes, Tao, and Oneness to form the Lower Dan Tian Jin Dan, which is called Ha Jin Dan.

大昌盛 Da Chang Sheng

"Da" means *greatest*. "Chang Sheng" means *flourishing*. "Da Chang Sheng" means *greatest flourishing*.

大昌盛金丹 Da Chang Sheng Jin Dan

"Da Chang Sheng Jin Dan" means *greatest flourishing golden light ball*. When you chant *Da Chang Sheng Jin Dan*, you are gathering Shen Qi Jing from Tao Source and Oneness, which are the Wu World, as well as from countless planets, stars, galaxies, and universes, which are the You World, to form Da Chang Sheng Jin Dan.

道賜盈福 Dao Ci Ying Fu

Dao is the Source and Ultimate Creator. "Ci" means *bestow*. "Ying" means *full* or *huge*. "Fu" means *good fortune and luck*. "Dao Ci Ying Fu" means *Tao Source bestows huge fortune, including Tao wisdom, health, longevity, happiness, good relationships, prosperity, luck, success, and more, to all aspects of life.*

行善積德 Xing Shan Ji De

"Xing Shan" means *offer kind service, including action, speech, and thought*. "Ji De" means *accumulate virtue*. "Xing Shan Ji De" means *offer kind service through action, speech, and thought to accumulate good virtue.*

道業昌盛 Dao Ye Chang Sheng

Dao is the Source and Ultimate Creator. "Ye" means *career*. "Chang Sheng" means *flourishes*. "Dao Ye Chang Sheng" means *Tao career flourishes*.

Deeper wisdom of Da Chang Sheng

Da Chang Sheng, greatest flourishing, is the seventh of the Ten Da natures of Tao. The four-line sacred Tao mantra of Da Chang Sheng is:

Qi Da Chang Sheng (七大昌盛)
Dao Ci Ying Fu (道賜盈福)
Xing Shan Ji De (行善積德)
Dao Ye Chang Sheng (道業昌盛)

The seventh of the Ten Da Tao Natures is greatest flourishing.
Tao Source bestows huge prosperity, luck, and success.
Offer kind service to accumulate virtue.
Tao career flourishes.

I would like to share a deep spiritual wisdom. The roots of a person's flourishing are positive karma from past lifetimes and personal effort in the current lifetime. If you have open spiritual channels, you may be able to see past lifetimes to understand this truth.

Financial success is due to positive karma from your
ancestors and you from past and current lifetimes
and great personal effort in the current lifetime.

Positive karma is accumulated through great service for humanity to make others healthier and happier and to transform their lives in a positive way. Karma in past lifetimes is recorded in one's Akashic Record book. It is the place to record a human being's and a soul's life. This positive karma can transform your financial flourishing in this lifetime and in future lifetimes. If you have great flourishing, congratulations. I wish for you to receive more. If you do not have great flourishing, how can you create it? The secret is to accumulate positive karma through Shen, Kou, and Yi (action, speech, and thought). To chant sacred Tao mantras is to accumulate positive karma. Your chanting can serve others and bring positive Shen Qi Jing of Tao to your life.

For example, a businessman in Los Angeles received one of my Tao calligraphies, "Dao Ye Chang Sheng (道業昌盛)," which means *Tao career flourishes*. He told me, "Since I received the 'Dao Ye Chang Sheng' Tao calligraphy, I have traced it and chanted for about ten minutes every day. Money comes without any effort. Everything is flowing. My company has grown from about a two-million-dollar business to sixty million dollars within two years. I absolutely give the credit to the 'Dao Ye Chang Sheng' calligraphy and mantra."

Because the "Dao Ye Chang Sheng" Tao mantra and calligraphy carry the Shen Qi Jing of Tao, his business received huge blessings from Tao Source. Chant these four sacred phrases to remove Shen Qi Jing blockages in your finances and business. Chant more to receive more benefits. Remember to use the heart to chant. Chant a minimum of ten minutes per day.

Practice

Apply the Four Power Techniques:

1. Body Power

Sit or stand as in the practice in chapter one. Remember to alternate sitting and standing positions in your practice for the best results.

2. Soul Power

Say *hello* to inner and outer souls:

> *Dear Tao Source,*
> *Dear Heaven and Mother Earth,*
> *Dear countless planets, stars, galaxies, and universes,*
> *Dear Bao Yuan Shou Yi Chang Shou Yong Sheng Dao Jing* (Tao Classic of
>   Longevity and Immortality),
> *I love you, honor you, and appreciate you.*
> *Please purify and remove my Shen Qi Jing blockages for healing, rejuvenation,*
>   *transformation of relationships and finances, increasing intelligence, and*
>   *flourishing in every aspect of my life.*
> *Please bless my spiritual practice of longevity and immortality and develop my*
>   *Lower Dan Tian and my Da Chang Sheng, greatest flourishing.*
> *Please bless me to achieve longevity and immortality.*
> *I am extremely honored and grateful.*
> *Thank you. Thank you. Thank you.*

3. Mind Power

Visualize a Tao Source Jin Dan (golden light ball) shining and rotating in your Lower Dan Tian (lower abdomen) and then shining throughout your entire body (figure 12). The light is cleansing Shen Qi Jing blockages in your Lower

Dan Tian and then getting stronger and brighter as your Jin Dan grows to the size of your body.

Figure 12. Lower Dan Tian Jin Dan Practice

4. Sound Power

Chant silently or aloud stanzas one through nine of the *Tao Classic of Longevity and Immortality* (from chapters one through nine), and then repeat the ninth stanza (in **bold** below) ten or more times in each practice session. Remember that best results are obtained by alternating silent and aloud (Yin Yang) chanting.

Hei Hei Hei
Hei Jin Dan
Da Guang Ming
Da Guang Ming Jin Dan
Wo Zai Dao Guang Zhong
Dao Guang Zai Wo Zhong
Tong Ti Tou Ming

Heng Heng Heng
Heng Jin Dan
Da Gan En
Da Gan En Jin Dan
Dao Sheng De Yang
Zai Pei Ci Hui
Dao En Yong Cun

Hong Hong Hong
Hong Jin Dan
Da Qian Bei
Da Qian Bei Jin Dan
Rou Ruo Bu Zheng
Chi Xu Jing Jin
Shi Qian Bei
Die Wan Zhang

Ah Ah Ah
Ah Jin Dan
Da Ai
Da Ai Jin Dan
Wu Tiao Jian Ai
Rong Hua Zai Nan
Xin Qing Shen Ming

Xi Xi Xi
Xi Jin Dan
Da Fu Wu
Da Fu Wu Jin Dan
Shi Wei Gong Pu
Wu Si Feng Xian
Shang Cheng Fa Men

Yi Yi Yi
Yi Jin Dan
Da Kuan Shu
Da Kuan Shu Jin Dan

Wo Yuan Liang Ni
Ni Yuan Liang Wo
Xiang Ai Ping An He Xie

Weng Weng Weng
Weng Jin Dan
Da Yuan Man
Da Yuan Man Jin Dan
Ling Xin Nao Shen Yuan Man
Ren Di Tian Dao Shen Xian Ti
Fu Wu Xiu Lian Cai Ke Pan

You You You
You Jin Dan
Da Ci Bei
Da Ci Bei Jin Dan
Yuan Li Zeng Qiang
Fu Wu Zhong Sheng
Gong De Wu Liang (1 time)

Ha Ha Ha
Ha Jin Dan
Da Chang Sheng
Da Chang Sheng Jin Dan
Dao Ci Ying Fu
Xing Shan Ji De
Dao Ye Chang Sheng (10 times)

Chapter Ten

Chinese	Pinyin	English Translation
吁吁吁	Yu Yu Yu	*Yu Yu Yu*
吁金丹	Yu Jin Dan	*Yu Golden Light Ball*
大和諧	Da He Xie	*Greatest Harmony*
大和諧金丹	Da He Xie Jin Dan	*Greatest Harmony Golden Light Ball*
三人同心	San Ren Tong Xin	*Three people join hearts together*
齊力斷金	Qi Li Duan Jin	*Their strength can cut through gold*
成功秘訣	Cheng Gong Mi Jue	*This is the Tao secret of success*

Sacred wisdom

吁吁吁 Yu Yu Yu (Wei Lü)
Yu is a sacred and secret Tao mantra to develop the Wei Lü, which is the tailbone area (figure 13).

吁金丹 Yu Jin Dan
When you chant *Yu Yu Yu, Yu Jin Dan*, you gather Shen Qi Jing in a secret way from countless planets, stars, galaxies, and universes, Tao, and Oneness to form the Wei Lü Jin Dan, which is called Yu Jin Dan.

大和諧 Da He Xie

"Da" means *greatest*. "He Xie" means *harmony*. "Da He Xie" means *greatest harmony*.

大和諧金丹 Da He Xie Jin Dan

"Da He Xie Jin Dan" means *greatest harmony golden light ball*. When you chant *Da He Xie Jin Dan*, you are gathering Shen Qi Jing from Tao Source and Oneness, which are the Wu World, as well as from countless planets, stars, galaxies, and universes, which are the You World, to form Da He Xie Jin Dan.

三人同心 San Ren Tong Xin

"San" means *three*. "Ren" means *human being*. "Tong" means *together*. "Xin" means *heart*. "San Ren Tong Xin" means *three people join their hearts together*.

齊力斷金 Qi Li Duan Jin

"Qi Li" means *the strength of three hearts joined together*. "Duan" means *cut*. "Jin" means *gold*. "Qi Li Duan Jin" means *the strength of three hearts joined together can cut through gold*.

成功秘訣 Cheng Gong Mi Jue

"Cheng Gong" means *success*. "Mi Jue" means *secret*. "Cheng Gong Mi Jue" means the *Tao secret wisdom of success*.

Deeper wisdom of Da He Xie

Da He Xie, greatest harmony, is the sixth of the Ten Da natures of Tao. It is another essential quality for true success in one's life.

The four-line sacred Tao mantra of Da He Xie is:

Liu Da He Xie (六大和諧)
San Ren Tong Xin (三人同行)
Qi Li Duan Jin (齊力斷金)
Cheng Gong Mi Jue (成功密訣)

The sixth of the Ten Da Tao Natures is greatest harmony.
Three people join hearts together.

Their strength can cut gold.
The secret of success.

To have a happy family, each family member must offer love, care, and compassion. Everyone needs to forgive each other and be in harmony. There is an ancient statement: Jia He Wan Shi Xing (家和萬事興). "Jia" means *family.* "He" means *harmony.* "Wan" means *ten thousand, countless.* "Shi" means *things.* "Xing" means *flourishing.* "Jia He Wan Shi Xing" means *in a harmonized family everything is flourishing.* Think about a very successful business. It must have a great harmonized team. Da He Xie represents great teamwork. Without Da He Xie, it is impossible to have great flourishing. To have great harmony, one must also have the other Ten Da natures.

The Ten Da are the natures of Tao, Buddha, saints, the Divine, countless planets, stars, galaxies, and universes, Mother Earth, and human beings. Da He Xie is great teamwork, which is the key for success. Let us love one another and join as one to create greatest success in every aspect of our lives.

Practice

Apply the Four Power Techniques:

1. Body Power

Sit or stand as in the practice in chapter one. Alternate sitting and standing positions in your practice in order to achieve the best results.

2. Soul Power

Say *hello* to inner and outer souls:

Dear Tao Source,
Dear Heaven and Mother Earth,
Dear countless planets, stars, galaxies, and universes,

Dear Bao Yuan Shou Yi Chang Shou Yong Sheng Dao Jing (Tao Classic of
    Longevity and Immortality),
I love you, honor you, and appreciate you.
*Please purify and remove my Shen Qi Jing blockages for healing, rejuvenation,
    transformation of relationships and finances, increasing intelligence, and
    flourishing in every aspect of my life.*
*Please bless my spiritual practice of longevity and immortality and develop my
    Wei Lü and my Da He Xie, greatest harmony.*
Please bless me to achieve longevity and immortality.
I am extremely honored and grateful.
Thank you. Thank you. Thank you.

3. Mind Power

Visualize a Tao Source Jin Dan (golden light ball) shining and rotating in the
Wei Lü area and then radiating light throughout your entire body (figure 13).
The light is cleansing Shen Qi Jing blockages in your Wei Lü area and then
getting stronger and brighter as you visualize the Jin Dan growing to the size
of your body.

Figure 13. Wei Lü Jin Dan Practice

4. Sound Power

Chant silently or aloud the first ten stanzas of the *Tao Classic of Longevity and Immortality* (from chapters one through ten), and then repeat the tenth stanza (in **bold** below) ten or more times in each practice session. Remember to alternate silent and aloud (Yin Yang) chanting.

Hei Hei Hei
Hei Jin Dan
Da Guang Ming
Da Guang Ming Jin Dan
Wo Zai Dao Guang Zhong
Dao Guang Zai Wo Zhong
Tong Ti Tou Ming

Heng Heng Heng
Heng Jin Dan
Da Gan En
Da Gan En Jin Dan
Dao Sheng De Yang
Zai Pei Ci Hui
Dao En Yong Cun

Hong Hong Hong
Hong Jin Dan
Da Qian Bei
Da Qian Bei Jin Dan
Rou Ruo Bu Zheng
Chi Xu Jing Jin
Shi Qian Bei
Die Wan Zhang

Ah Ah Ah
Ah Jin Dan
Da Ai
Da Ai Jin Dan
Wu Tiao Jian Ai

Rong Hua Zai Nan
Xin Qing Shen Ming

Xi Xi Xi
Xi Jin Dan
Da Fu Wu
Da Fu Wu Jin Dan
Shi Wei Gong Pu
Wu Si Feng Xian
Shang Cheng Fa Men

Yi Yi Yi
Yi Jin Dan
Da Kuan Shu
Da Kuan Shu Jin Dan
Wo Yuan Liang Ni
Ni Yuan Liang Wo
Xiang Ai Ping An He Xie

Weng Weng Weng
Weng Jin Dan
Da Yuan Man
Da Yuan Man Jin Dan
Ling Xin Nao Shen Yuan Man
Ren Di Tian Dao Shen Xian Ti
Fu Wu Xiu Lian Cai Ke Pan

You You You
You Jin Dan
Da Ci Bei
Da Ci Bei Jin Dan
Yuan Li Zeng Qiang
Fu Wu Zhong Sheng
Gong De Wu Liang

Ha Ha Ha
Ha Jin Dan

Da Chang Sheng
Da Chang Sheng Jin Dan
Dao Ci Ying Fu
Xing Shan Ji De
Dao Ye Chang Sheng (1 time)

Yu Yu Yu
Yu Jin Dan
Da He Xie
Da He Xie Jin Dan
San Ren Tong Xin
Qi Li Duan Jin
Cheng Gong Mi Jue (10 times)

Chapter Eleven

Chinese	Pinyin	English Translation
嘿哼哄啊唏嚱嗡呦	Hei Heng Hong Ah Xi Yi Weng You	*Divine and Tao Source Qi Channel for healing*
呦嗡嚱唏啊哄哼嘿	You Weng Yi Xi Ah Hong Heng Hei	*Divine and Tao Source Jing Channel for rejuvenation and longevity*
嗡嘿哄呦	Weng Hei Hong You	*Divine and Tao Source Shen Channel for immortality*
光感谦爱服宽圆慈	Guang Gan Qian Ai Fu Kuan Yuan Ci	*Tao Source Eight Da Natures Qi Channel for healing*
慈圆宽服爱谦感光	Ci Yuan Kuan Fu Ai Qian Gan Guang	*Tao Source Eight Da Natures Jing Channel for rejuvenation and longevity*
圆光谦慈	Yuan Guang Qian Ci	*Tao Source Four Da Natures Shen Channel for immortality*

Sacred wisdom

Now we are going to purify and remove Shen Qi Jing blockages within the most important Shen Qi Jing channels in the body.

Qi Channel
The Qi Channel is the major energy channel in a human body. It starts at the Hui Yin acupuncture point on the perineum, at the base of the first Soul House. It moves up through the second, third, fourth, fifth, and sixth Soul Houses to

the Bai Hui acupuncture point on the crown of the head and at the base of the seventh Soul House. From there, the Qi Channel turns down through the Wai Jiao back to the first Soul House. The Wai Jiao is the space in front of the spinal column. It is the biggest space in the body (figure 14, top). The Qi Channel is the key for healing all sickness.

The sacred and secret mantra of the Qi Channel is:

嘿哼哄啊唏噎嗡呦 Hei Heng Hong Ah Xi Yi Weng You.

The Qi Channel is the pathway of the seven Soul Houses, San Jiao, and Wai Jiao. Hei Heng Hong Ah Xi Yi Weng You assembles all the secret mantras on the pathway. Therefore, this mantra not only purifies and removes Shen Qi Jing blockages from the Qi Channel, it also further removes Shen Qi Jing blockages from and strengthens each Soul House, the San Jiao, and Wai Jiao. The more you chant, the more benefits you will receive on this pathway.

Jing Channel

The Jing Channel (figure 14, middle) moves in the reverse direction of the Qi Channel. The Jing Channel also starts at the Hui Yin acupuncture point on the perineum but moves up and back to the tailbone, goes through an invisible hole in the tailbone, and connects with the spinal cord. It then goes up through the spinal cord to the brain and up to the Bai Hui acupuncture point on the crown of the head. Next, it flows down the central channel through the sixth, fifth, fourth, third, second, and first Soul Houses, returning to the Hui Yin acupuncture point. The Jing Channel is the key for rejuvenation and longevity.

The sacred and secret mantra for the Jing Channel is:

呦嗡噎唏啊哄哼嘿 You Weng Yi Xi Ah Hong Heng Hei.

When you chant *You Weng Yi Xi Ah Hong Heng Hei*, you are cleansing the sacred pathway for rejuvenation and longevity. Because the Qi Channel and Jing Channel have similar pathways, the sacred Jing Channel mantra can also further remove Shen Qi Jing blockages from and strengthen each Soul House, the San Jiao, and Wai Jiao. Therefore, chant more to receive more benefits in these vital spaces of the body.

Four ancient sacred phrases explain the Jing Channel:

Lian Jing Hua Qi (煉精化氣)
Lian Qi Hua Shen (煉氣化神)
Lian Shen Huan Xu (煉神還虛)
Lian Xu Huan Dao (煉虛還道)

"Lian" and "Hua" both mean *transform*. "Jing" means *matter*. "Qi" means *energy*. "Lian Jing Hua Qi" means *transform Jing to Qi*. According to ancient Tao wisdom and traditional Chinese medicine, the kidneys produce Jing, the essence of matter. This Jing flows to the tailbone through an invisible hole up to the spinal cord and transforms to Qi, energy, which has higher frequency. "Lian Qi Hua Shen" means *transform Qi to Shen*. Qi flows up inside the spinal column to the brain. Before it reaches the brain, Lian Qi Hua Shen. The energy purifies and is uplifted to the Shen level, which includes soul, heart, and mind. "Lian Shen Huan Xu" means *return soul, heart, and mind to emptiness*. The Shen flows from the brain down to the heart, where soul, heart, and mind can be purified and enlightened to the empty state. "Lian Xu Huan Dao" means *return emptiness to Tao*. "Xu" is *emptiness*. "Dao" is *ultimate emptiness*. Tao is in the Zhong, which is essentially the back half of the lower abdomen. I will explain further knowledge and wisdom of Zhong in chapter fourteen. You Weng Yi Xi Ah Hong Heng Hei is the simplified version of this ancient four-phrase Tao wisdom and practice.

Shen Channel

The Shen Channel (figure 14, bottom) is the immortality pathway. It starts simultaneously at the Bai Hui acupuncture point at the base of the seventh Soul House with the secret mantra *Weng* and at the Hui Yin acupuncture point at the base of the first Soul House with the secret mantra *Hei*. From these two points, the Shen Channel flows through the central channel down from the seventh Soul House and up from the first Soul House, to meet behind the navel in the third Soul House with the secret mantra *Hong*. From there, it flows straight back to the Ming Men acupuncture point where it splits, with one part going up through the Wai Jiao, returning to the Bai Hui acupuncture point, and the other part going down through the Wai Jiao, returning to the Hui Yin acupuncture point, all with the secret mantra *You*.

Therefore, the sacred and secret mantra of the Shen Channel is: 嗡嘿哄呦 Weng Hei Hong You.

The secret mantra *Weng* connects with Heaven. The secret mantra *Hei* connects with Mother Earth. The secret mantra *Hong* connects with a human being. The secret mantra *You* connects with Tao. When you chant *Weng Hei Hong You*, the Shen Qi Jing of Heaven, Mother Earth, human being, and Tao will join as one. Therefore, this sacred mantra is to practice and achieve immortality.

光感謙愛服寬圓慈 Guang Gan Qian Ai Fu Kuan Yuan Ci

This mantra is for practicing and purifying the Qi Channel further to heal all sickness. The functions are the same as with the mantra *Hei Heng Hong Ah Xi Yi Weng You*. *Hei Heng Hong Ah Xi Yi Weng You* is the secret Tao mantra with sacred meaning. *Guang Gan Qian Ai Fu Kuan Yuan Ci* is the Tao mantra that incorporates eight of the Ten Da Tao Natures.

"Guang" refers to Da Guang Ming, which is greatest light. It is the sacred mantra for the first Soul House.

"Gan" refers to Da Gan En, which is greatest gratitude. It is the sacred mantra for the second Soul House.

"Qian" refers to Da Qian Bei, which is greatest humility. It is the sacred mantra for the third Soul House.

"Ai" refers to Da Ai, which is greatest love. It is the sacred mantra for the fourth Soul House.

"Fu" refers to Da Fu Wu, which is greatest service. It is the sacred mantra for the fifth Soul House.

"Kuan" refers to Da Kuan Shu, which is greatest forgiveness. It is the sacred mantra for the sixth Soul House.

"Yuan" refers to Da Yuan Man, which is greatest enlightenment. It is the sacred mantra for the seventh Soul House.

"Ci" refers to Da Ci Bei, which is greatest compassion. It is the sacred mantra for the Ming Men point, Kundalini, and Wai Jiao.

Each Da has unique power and abilities to transform all life. Applying one Da is powerful. Applying eight Da together is beyond powerful. The mantra *Guang Gan Qian Ai Fu Kuan Yuan Ci* is the key for healing all sickness of the body.

慈圓寬服愛謙感光 Ci Yuan Kuan Fu Ai Qian Gan Guang

Similarly, *Ci Yuan Kuan Fu Ai Qian Gan Guang* is the mantra of the Jing Channel with eight of the Ten Da Tao Natures. The functions are the same as in the mantra *You Weng Yi Xi Ah Hong Heng Hei*. Both are powerful mantras for rejuvenation and longevity. It is the reverse order of the Qi Channel mantra *Guang Gan Qian Ai Fu Kuan Yuan Ci*. Chant *Ci Yuan Kuan Fu Ai Qian Gan Guang* as much as you can to receive immeasurable benefits for your Jing Channel.

圓光謙慈 Yuan Guang Qian Ci

Yuan Guang Qian Ci is the mantra for the Shen Channel with four of the Ten Da Tao Natures. The Shen Channel is the Ren Di Tian Dao He Yi (人地天道

合一) Channel, where human being, Mother Earth, Heaven, and Tao join as one. Therefore, the Shen Channel is the Tao Channel. To practice the mantras *Weng Hei Hong You* and *Yuan Guang Qian Ci* is to practice and achieve immortality. Chant *Yuan Guang Qian Ci* as much as you can. The benefits are beyond comprehension.

Practice

Apply the Four Power Techniques:

1. Body Power

Sit or stand as in the practice in chapter one. Alternate sitting and standing positions in your practice.

2. Soul Power

Say *hello* to inner and outer souls:

> *Dear Tao Source,*
> *Dear Heaven and Mother Earth,*
> *Dear countless planets, stars, galaxies, and universes,*
> *Dear Bao Yuan Shou Yi Chang Shou Yong Sheng Dao Jing* (Tao Classic of Longevity and Immortality),
> *I love you, honor you, and appreciate you.*
> *Please purify and remove my Shen Qi Jing blockages for healing, rejuvenation, transformation of relationships and finances, increasing intelligence, and flourishing in every aspect of my life.*
> *Please bless my spiritual practice of longevity and immortality and develop my Shen, Qi, and Jing Channels.*
> *Please bless me to achieve longevity and immortality.*
> *I am extremely honored and grateful.*
> *Thank you. Thank you. Thank you.*

3. Mind Power

When you chant *Hei Heng Hong Ah Xi Yi Weng You* and *Guang Gan Qian Ai Fu Kuan Yuan Ci*, visualize a golden light ball moving from the first Soul House up to the seventh Soul House, then down through the Wai Jiao back to the first Soul House to form a circle (figure 14, top).

When you chant *You Weng Yi Xi Ah Hong Heng Hei* and *Ci Yuan Kuan Fu Ai Qian Gan Guang*, visualize a golden light ball moving from the first Soul House to the tailbone, then up the spinal cord to the Bai Hui acupuncture point at the top of the head. From there, it flows down the center of the body through all seven Soul Houses back to the Hui Yin acupuncture point (figure 14, middle).

When you chant *Weng Hei Hong You* and *Yuan Guang Qian Ci*, visualize two golden light balls, one in the seventh Soul House and one in the first Soul House. They move through the center of the body and join as one in the third Soul House. The one ball moves straight back to the Ming Men acupuncture point, where it subdivides into two balls that travel respectively up and down through the Wai Jiao, returning to the Bai Hui and Hui Yin acupuncture points (figure 14, bottom).

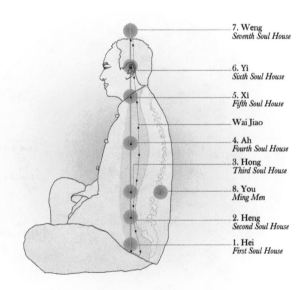

7. Weng
Seventh Soul House

6. Yi
Sixth Soul House

5. Xi
Fifth Soul House

Wai Jiao

4. Ah
Fourth Soul House

3. Hong
Third Soul House

8. You
Ming Men

2. Heng
Second Soul House

1. Hei
First Soul House

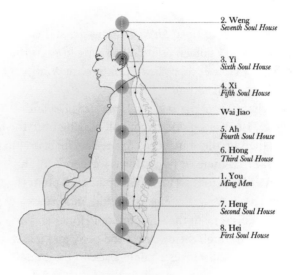

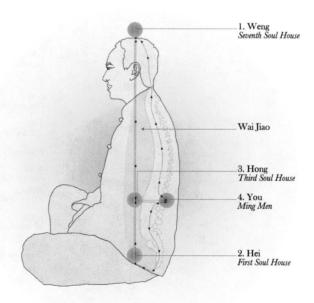

Figure 14. Qi (top), Jing (middle), Shen (bottom) Channels Practice

4. Sound Power

Chant silently or aloud stanzas one through eleven of the *Tao Classic of Longevity and Immortality* (from chapters one through eleven), and then repeat the eleventh stanza (in **bold** below) ten or more times in each practice session. Remember to alternate silent and aloud (Yin Yang) chanting.

Hei Hei Hei
Hei Jin Dan
Da Guang Ming
Da Guang Ming Jin Dan
Wo Zai Dao Guang Zhong
Dao Guang Zai Wo Zhong
Tong Ti Tou Ming

Heng Heng Heng
Heng Jin Dan
Da Gan En
Da Gan En Jin Dan
Dao Sheng De Yang
Zai Pei Ci Hui
Dao En Yong Cun

Hong Hong Hong
Hong Jin Dan
Da Qian Bei
Da Qian Bei Jin Dan
Rou Ruo Bu Zheng
Chi Xu Jing Jin
Shi Qian Bei
Die Wan Zhang

Ah Ah Ah
Ah Jin Dan
Da Ai
Da Ai Jin Dan
Wu Tiao Jian Ai

Rong Hua Zai Nan
Xin Qing Shen Ming

Xi Xi Xi
Xi Jin Dan
Da Fu Wu
Da Fu Wu Jin Dan
Shi Wei Gong Pu
Wu Si Feng Xian
Shang Cheng Fa Men

Yi Yi Yi
Yi Jin Dan
Da Kuan Shu
Da Kuan Shu Jin Dan
Wo Yuan Liang Ni
Ni Yuan Liang Wo
Xiang Ai Ping An He Xie

Weng Weng Weng
Weng Jin Dan
Da Yuan Man
Da Yuan Man Jin Dan
Ling Xin Nao Shen Yuan Man
Ren Di Tian Dao Shen Xian Ti
Fu Wu Xiu Lian Cai Ke Pan

You You You
You Jin Dan
Da Ci Bei
Da Ci Bei Jin Dan
Yuan Li Zeng Qiang
Fu Wu Zhong Sheng
Gong De Wu Liang

Ha Ha Ha
Ha Jin Dan

Da Chang Sheng
Da Chang Sheng Jin Dan
Dao Ci Ying Fu
Xing Shan Ji De
Dao Ye Chang Sheng

Yu Yu Yu
Yu Jin Dan
Da He Xie
Da He Xie Jin Dan
San Ren Tong Xin
Qi Li Duan Jin
Cheng Gong Mi Jue (1 time)

Hei Heng Hong Ah Xi Yi Weng You
Hei Heng Hong Ah Xi Yi Weng You
Hei Heng Hong Ah Xi Yi Weng You
Hei Heng Hong Ah Xi Yi Weng You

You Weng Yi Xi Ah Hong Heng Hei
You Weng Yi Xi Ah Hong Heng Hei
You Weng Yi Xi Ah Hong Heng Hei
You Weng Yi Xi Ah Hong Heng Hei

Weng Hei Hong You
Weng Hei Hong You
Weng Hei Hong You
Weng Hei Hong You

Guang Gan Qian Ai Fu Kuan Yuan Ci
Guang Gan Qian Ai Fu Kuan Yuan Ci
Guang Gan Qian Ai Fu Kuan Yuan Ci
Guang Gan Qian Ai Fu Kuan Yuan Ci

Ci Yuan Kuan Fu Ai Qian Gan Guang
Ci Yuan Kuan Fu Ai Qian Gan Guang
Ci Yuan Kuan Fu Ai Qian Gan Guang

Ci Yuan Kuan Fu Ai Qian Gan Guang

Yuan Guang Qian Ci
Yuan Guang Qian Ci
Yuan Guang Qian Ci
Yuan Guang Qian Ci (10 times)

Part II
Tao Body

TO FULFILL THE leaky body, there are two steps: One is to purify the spaces. I have shared the secret wisdom for this in step I. The other part is to purify the organs. According to ancient Tao wisdom and traditional Chinese medicine, the human body is made up of five basic elements (figure 15). Each element is led by an authority organ that is in charge of the entire element.[7] The Five Elements, their authority organs, and more are:

Wood element
The Wood element includes the liver, gallbladder, eyes, tendons, and, in the emotional body, anger. The authority organ of the Wood element is the liver. To purify and to remove Shen Qi Jing blockages from the liver will directly benefit everything within the Wood element.

---

7 Dr. and Master Zhi Gang Sha, *Soul Healing Miracles: Ancient and New Sacred Wisdom, Knowledge, and Practical Techniques for Healing the Spiritual, Mental, Emotional, and Physical Bodies,* Dallas/Toronto: BenBella Books/Heaven's Library, 2013.

Fire element

The Fire element includes the heart, small intestine, tongue, blood vessels, and, in the emotional body, depression and anxiety. The authority organ of the Fire element is the heart. To purify and to remove Shen Qi Jing blockages from the heart will directly benefit everything within the Fire element.

Earth element

The Earth element includes the spleen, stomach, mouth, lips, teeth, gums, muscles, and, in the emotional body, worry. The authority organ of the Earth element is the spleen. To purify and to remove Shen Qi Jing blockages from the spleen will directly benefit everything within the Earth element.

Metal element

The Metal element includes the lungs, large intestine, nose, skin, and, in the emotional body, sadness and grief. The authority organ of the Metal element is the lungs. To purify and to remove Shen Qi Jing blockages from the lungs will directly benefit everything within the Metal element.

Water element

The Water element includes the kidneys, urinary bladder, ears, bones, and, in the emotional body, fear. The authority organ of the Water element is the kidneys. To purify and to remove Shen Qi Jing blockages from the kidneys will directly benefit everything within the Water element.

In part II (chapters twelve and thirteen), I will share the sacred mantras for the authority organs of the Five Elements and more to fulfill your leaky body and to purify, develop, and enlighten your physical body.

Chapter Twelve

Chinese	Pinyin	English Translation
嘘嘘嘘	Xu Xu Xu	*Xu Xu Xu*
嘘金丹	Xu Jin Dan	*Xu Golden Light Ball*
舒肝金丹	Shu Gan Jin Dan	*Smooth Liver Function Golden Light Ball*
啊啊啊	Ah Ah Ah	*Ah Ah Ah*
啊金丹	Ah Jin Dan	*Ah Golden Light Ball*
養心金丹	Yang Xin Jin Dan	*Nourish Heart Golden Light Ball*
呼呼呼	Hu Hu Hu	*Hu Hu Hu*
呼金丹	Hu Jin Dan	*Hu Golden Light Ball*
健脾金丹	Jian Pi Jin Dan	*Strengthen Spleen Golden Light Ball*
呬呬呬	Si Si Si	*Si Si Si*
呬金丹	Si Jin Dan	*Si Golden Light Ball*
宣肺金丹	Xuan Fei Jin Dan	*Facilitate Lung Function Golden Light Ball*
吹吹吹	Chui Chui Chui	*Chui Chui Chui*
吹金丹	Chui Jin Dan	*Chui Golden Light Ball*

| 壯腎金丹 | Zhuang Shen Jin Dan | *Reinforce Kidneys Golden Light Ball* |

Sacred wisdom

噓噓噓 Xu Xu Xu (Wood element)
Xu is a sacred and secret Tao mantra to purify and remove Shen Qi Jing blockages in the liver and the entire Wood element, including the gallbladder, eyes, tendons, and, in the emotional body, anger.

噓金丹 Xu Jin Dan
When you chant *Xu Xu Xu, Xu Jin Dan*, you gather Shen Qi Jing in a secret way from countless planets, stars, galaxies, and universes, Tao, and Oneness to remove Shen Qi Jing blockages from the liver and the entire Wood element in order to create a Xu Jin Dan. It is a mantra with sacred messages.

舒肝金丹 Shu Gan Jin Dan
"Shu Gan" means *smooth and adjust liver functions*. Shu Gan Jin Dan is a golden light ball to smooth and adjust the liver. When you chant *Shu Gan Jin Dan*, you gather Shen Qi Jing from countless planets, stars, galaxies, and universes, Tao, and Oneness to form the Shu Gan Jin Dan.

啊啊啊 Ah Ah Ah (Fire element)
Ah is a sacred and secret Tao mantra to purify and remove Shen Qi Jing blockages from the heart and the entire Fire element, including the small intestine, tongue, blood vessels, and, in the emotional body, depression and anxiety.

啊金丹 Ah Jin Dan
When you chant *Ah Ah Ah, Ah Jin Dan*, you gather Shen Qi Jing from countless planets, stars, galaxies, and universes, Tao, and Oneness to form the heart and Fire element Jin Dan, which is called Ah Jin Dan. It is a mantra with sacred messages.

養心金丹 Yang Xin Jin Dan
"Yang Xin" means *nourish the heart*. Yang Xin Jin Dan is a golden light ball to nourish the heart. When you chant *Yang Xin Jin Dan*, you gather Shen Qi Jing

from countless planets, stars, galaxies, and universes, Tao, and Oneness to form the Yang Xin Jin Dan.

呼呼呼 Hu Hu Hu (Earth element)
Hu is a sacred and secret Tao mantra to purify and remove Shen Qi Jing blockages from the spleen and the entire Earth element, including the stomach, mouth, lips, gums, teeth, muscles, and, in the emotional body, worry.

呼金丹 Hu Jin Dan
When you chant *Hu Hu Hu, Hu Jin Dan*, you gather Shen Qi Jing in a secret way from countless planets, stars, galaxies, and universes, Tao, and Oneness to form the spleen and Earth element Jin Dan, which is called Hu Jin Dan. It is a mantra with sacred messages.

健脾金丹 Jian Pi Jin Dan
"Jian Pi" means *strengthen the spleen*. Jian Pi Jin Dan is a golden light ball to strengthen and develop the spleen. When you chant *Jian Pi Jin Dan*, you gather Shen Qi Jing from countless planets, stars, galaxies, and universes, Tao, and Oneness to form the Jian Pi Jin Dan.

呬呬呬 Si Si Si (Metal element)
Si is a sacred and secret Tao mantra to purify and remove Shen Qi Jing blockages in the lungs and the entire Metal element, including the large intestine, nose, skin, and, in the emotional body, sadness and grief.

呬金丹 Si Jin Dan
When you chant *Si Si Si, Si Jin Dan*, you gather Shen Qi Jing in a secret way from countless planets, stars, galaxies, and universes, Tao, and Oneness to form the lungs and Metal element Jin Dan, which is called Si Jin Dan. It is a mantra with sacred messages.

宣肺金丹 Xuan Fei Jin Dan
"Xuan Fei" means *disperse lung Qi*. Xuan Fei Jin Dan is a golden light ball to disperse lung Qi and to strengthen the entire Metal element. When you chant *Xuan Fei Jin Dan*, you gather Shen Qi Jing from countless planets, stars, galaxies, and universes, Tao, and Oneness to form the Xuan Fei Jin Dan.

吹吹吹 Chui Chui Chui (Water element)

Chui is a sacred and secret Tao mantra to purify and remove Shen Qi Jing blockages from the kidneys and the entire Water element, including the urinary bladder, ears, bones, and, in the emotional body, fear.

吹金丹 Chui Jin Dan

When you chant *Chui Chui Chui, Chui Jin Dan*, you gather Shen Qi Jing in a secret way from countless planets, stars, galaxies, and universes, Tao, and Oneness to form the kidneys and Water element Jin Dan, which is called Chui Jin Dan. It is a mantra with sacred messages.

壯腎金丹 Zhuang Shen Jin Dan

"Zhuang Shen" means *strengthen the kidneys.* Zhuang Shen Jin Dan is a golden light ball to strengthen and develop the kidneys. When you chant *Zhuang Shen Jin Dan*, you gather the Shen Qi Jing from countless planets, stars, galaxies, and universes, Tao, and Oneness to form the Zhuang Shen Jin Dan.

In summary, to reach longevity and immortality, the first step is Bu Lou Zhu Ji, fulfill the leaky body and build a foundation. To clear Shen Qi Jing blockages in the spaces (part I, chapters one through twelve) is the sacred way to heal and prevent all sickness. To purify and heal major organs is also a very important practice to heal all sickness. Both practices are equally important for the longevity and immortality journey.

Practice

Apply the Four Power Techniques:

1. Body Power

Sit or stand as in the practice in chapter one. Alternate sitting and standing positions in your practice for the best results.

2. Soul Power

Say *hello* to inner and outer souls:

> *Dear Tao Source,*
> *Dear Heaven and Mother Earth,*
> *Dear countless planets, stars, galaxies, and universes,*
> *Dear Bao Yuan Shou Yi Chang Shou Yong Sheng Dao Jing* (Tao Classic of
>   Longevity and Immortality),
> *I love you, honor you, and appreciate you.*
> *Please purify and remove my Shen Qi Jing blockages for healing, rejuvenation,*
>   *transformation of relationships and finances, increasing intelligence, and*
>   *flourishing in every aspect of my life.*
> *Please bless my spiritual practice of longevity and immortality and purify and*
>   *develop my Five Elements.*
> *Please bless me to achieve longevity and immortality.*
> *I am extremely honored and grateful.*
> *Thank you. Thank you. Thank you.*

3. Mind Power

The visualizing practices below will guide you step by step to develop the Five Elements' Jin Dans (figure 15) and enlighten each of them.

When you chant *Xu Xu Xu, Xu Jin Dan, Shu Gan Jin Dan*, visualize a golden light ball shining in and around your liver to heal, strengthen, and prevent sickness in the liver and the rest of the Wood element, including the gallbladder, eyes, tendons, and, in the emotional body, anger. This could also enlighten your liver.

When you chant *Ah Ah Ah, Ah Jin Dan, Yang Xin Jin Dan*, visualize a golden light ball shining in and around your heart to heal, strengthen, and prevent sickness in the heart and the rest of the Fire element, including the small intestine, tongue, blood vessels, and, in the emotional body, depression and anxiety. This could also enlighten your heart.

When you chant *Hu Hu Hu, Hu Jin Dan, Jian Pi Jin Dan*, visualize a golden light ball shining in and around your spleen to heal, strengthen, and prevent sickness in the spleen and the rest of the Earth element, including the stomach, mouth, lips, teeth, gums, muscles, and, in the emotional body, worry. This could also enlighten your spleen.

When you chant *Si Si Si, Si Jin Dan, Xuan Fei Jin Dan*, visualize a golden light ball shining in and around your lungs to heal, strengthen, and prevent sickness in the lungs and the rest of the Metal element, including the large intestine, skin, nose, and, in the emotional body, sadness and grief. This could also enlighten your lungs.

When you chant *Chui Chui Chui, Chui Jin Dan, Zhuang Shen Jin Dan*, visualize a golden light ball shining in and around your kidneys to heal, strengthen, and prevent sickness in the kidneys and the rest of the Water element, including the urinary bladder, ears, bones, and, in the emotional body, fear. This could also enlighten your kidneys.

Figure 15. Five Elements Jin Dan Practice

4. Sound Power

Chant silently or aloud the first twelve stanzas of the *Tao Classic of Longevity and Immortality* (from chapters one through twelve), and then repeat the twelfth stanza (in **bold** below) ten or more times in each practice session. Remember to alternate silent and aloud (Yin Yang) chanting for best results.

Hei Hei Hei
Hei Jin Dan
Da Guang Ming
Da Guang Ming Jin Dan
Wo Zai Dao Guang Zhong
Dao Guang Zai Wo Zhong
Tong Ti Tou Ming

Heng Heng Heng
Heng Jin Dan
Da Gan En
Da Gan En Jin Dan
Dao Sheng De Yang
Zai Pei Ci Hui
Dao En Yong Cun

Hong Hong Hong
Hong Jin Dan
Da Qian Bei
Da Qian Bei Jin Dan
Rou Ruo Bu Zheng
Chi Xu Jing Jin
Shi Qian Bei
Die Wan Zhang

Ah Ah Ah
Ah Jin Dan
Da Ai
Da Ai Jin Dan
Wu Tiao Jian Ai

Rong Hua Zai Nan
Xin Qing Shen Ming

Xi Xi Xi
Xi Jin Dan
Da Fu Wu
Da Fu Wu Jin Dan
Shi Wei Gong Pu
Wu Si Feng Xian
Shang Cheng Fa Men

Yi Yi Yi
Yi Jin Dan
Da Kuan Shu
Da Kuan Shu Jin Dan
Wo Yuan Liang Ni
Ni Yuan Liang Wo
Xiang Ai Ping An He Xie

Weng Weng Weng
Weng Jin Dan
Da Yuan Man
Da Yuan Man Jin Dan
Ling Xin Nao Shen Yuan Man
Ren Di Tian Dao Shen Xian Ti
Fu Wu Xiu Lian Cai Ke Pan

You You You
You Jin Dan
Da Ci Bei
Da Ci Bei Jin Dan
Yuan Li Zeng Qiang
Fu Wu Zhong Sheng
Gong De Wu Liang

Ha Ha Ha
Ha Jin Dan

Da Chang Sheng
Da Chang Sheng Jin Dan
Dao Ci Ying Fu
Xing Shan Ji De
Dao Ye Chang Sheng

Yu Yu Yu
Yu Jin Dan
Da He Xie
Da He Xie Jin Dan
San Ren Tong Xin
Qi Li Duan Jin
Cheng Gong Mi Jue

Hei Heng Hong Ah Xi Yi Weng You
Hei Heng Hong Ah Xi Yi Weng You
Hei Heng Hong Ah Xi Yi Weng You
Hei Heng Hong Ah Xi Yi Weng You

You Weng Yi Xi Ah Hong Heng Hei
You Weng Yi Xi Ah Hong Heng Hei
You Weng Yi Xi Ah Hong Heng Hei
You Weng Yi Xi Ah Hong Heng Hei

Weng Hei Hong You
Weng Hei Hong You
Weng Hei Hong You
Weng Hei Hong You

Guang Gan Qian Ai Fu Kuan Yuan Ci
Guang Gan Qian Ai Fu Kuan Yuan Ci
Guang Gan Qian Ai Fu Kuan Yuan Ci
Guang Gan Qian Ai Fu Kuan Yuan Ci

Ci Yuan Kuan Fu Ai Qian Gan Guang
Ci Yuan Kuan Fu Ai Qian Gan Guang
Ci Yuan Kuan Fu Ai Qian Gan Guang

Ci Yuan Kuan Fu Ai Qian Gan Guang

Yuan Guang Qian Ci
Yuan Guang Qian Ci
Yuan Guang Qian Ci
Yuan Guang Qian Ci (1 time)

Xu Xu Xu
Xu Jin Dan
Shu Gan Jin Dan

Ah Ah Ah
Ah Jin Dan
Yang Xin Jin Dan

Hu Hu Hu
Hu Jin Dan
Jian Pi Jin Dan

Si Si Si
Si Jin Dan
Xuan Fei Jin Dan

Chui Chui Chui
Chui Jin Dan
Zhuang Shen Jin Dan (10 times)

Chapter Thirteen

Chinese	Pinyin	English Translation
目不妄視	Mu Bu Wang Shi	*Do not look at impure things*
耳不妄聽	Er Bu Wang Ting	*Do not listen to impure sounds*
口不妄言	Kou Bu Wang Yan	*Do not speak impure words*
外三寶不漏	Wai San Bao Bu Lou	*Three external treasures (eyes, ears, mouth) do not leak*
不視安神於心	Bu Shi An Shen Yu Xin	*Do not look at impure things to keep your Shen (soul, heart, mind) pure and peaceful in your heart*
不聽蓄精於腎	Bu Ting Xu Jing Yu Shen	*Do not listen to impure sounds to increase the pure essence of matter in your kidneys*
不言孕氣丹田	Bu Yan Yun Qi Dan Tian	*Do not speak impure words to increase the pure energy in your Lower Dan Tian*
內三寶自合	Nei San Bao Zi He	*Three internal treasures (Jing [matter], Qi [energy], Shen [soul, heart, mind]) join together as one*
外不漏	Wai Bu Lou	*Three external treasures do not leak*

內自合	Nei Zi He	*Three internal treasures (Shen Qi Jing) join as one*
通天達地	Tong Tian Da Di	*Connect through Heaven and Mother Earth as one*
逍遙道中	Xiao Yao Dao Zhong	*Meld with Tao and flow freely in the Tao*

Sacred wisdom

目不妄視 Mu Bu Wang Shi

"Mu" means *eye*. "Bu" means *not*. "Wang" means *random*. "Shi" means *to look at*. "Mu Bu Wang Shi" means *do not look at things randomly to avoid seeing impure and polluted things*. According to ancient spiritual teaching, the eyes are the windows of the heart and the soul. When you look at impure and polluted things, your heart and soul could be infected and even get lost. This is the first step of purification, to stop being polluted. Then your heart and soul will become purer and purer with your Xiu Lian (purification practice).

耳不妄聽 Er Bu Wang Ting

"Er" means *ear*. "Ting" means *to listen*. "Er Bu Wang Ting" means *do not listen to sounds randomly to avoid hearing impure and polluted sounds*. Human language carries positive and negative power. The ears and hearing connect with the kidneys through the Water element. The kidneys produce Jing (the essence of matter). Therefore, to hear negative sounds, words, and messages will cause you to leak Jing. This is how hearing and matter (Jing) are connected.

口不妄言 Kou Bu Wang Yan

"Kou" means *mouth*. "Yan" means *to speak*. "Kou Bu Wang Yan" means *do not speak carelessly with impure or polluted words*. When you speak, you use energy (Qi). If you speak negative words, you could seriously lose your Qi. When we speak negative words, it is usually with negative emotions, such as anger, upset, or overexcitement. This could easily cause significant loss of Qi. It is why we can feel exhausted after such outbursts.

Furthermore, an ancient spiritual teaching says, "Huo Cong Kou Chu (禍 從口出)," which means *disasters and misfortune come from the mouth*. Speaking too much, especially speaking carelessly and negatively, is a fast way to create negative karma and burn your virtue. Virtue is spiritual currency to bless every aspect of your life. To lose virtue is to harm your spiritual life and physical life.

外三寶不漏 Wai San Bao Bu Lou

"Wai" means *external*. "Wai San Bao" means *three external treasures*, which are eyes, ears, and mouth. "Bu Lou" means *do not leak*. "Wai San Bao Bu Lou" means *do not leak your Jing Qi Shen through your eyes (seeing), ears (hearing), and mouth (speaking)*.

不視安神於心 Bu Shi An Shen Yu Xin

"Bu Shi" means *do not look at or see*. "An Shen" means *peaceful Shen (soul, heart, mind)*. "Yu Xin" means *in the heart*. "Bu Shi An Shen Yu Xin" means *do not look at impure things to keep your Shen (soul, heart, mind) pure and peaceful in your heart*. This is the Tao teaching on how to develop peace and nourish your Shen.

不聽蓄精於腎 Bu Ting Xu Jing Yu Shen

"Bu Ting" means *do not hear*. "Xu Jing" means *to accumulate Jing (matter)*. This character "Shen" means *kidney*. "Bu Ting Xu Jing Yu Shen" means *do not hear impure sounds to accumulate Jing in your kidneys*. This is the Tao teaching on how to gather and accumulate your Jing.

不言孕氣丹田 Bu Yan Yun Qi Dan Tian

"Bu Yan" means *do not speak*. "Yun Qi" means *to increase Qi (energy)*. "Dan Tian" is the Lower Dan Tian, a foundational energy center in the lower abdomen. "Bu Yan Yun Qi Dan Tian" means *do not speak too much, especially by speaking impure words, in order to increase the pure energy in your Lower Dan Tian*. This is the Tao teaching on how to increase and accumulate your Qi.

內三寶自合 Nei San Bao Zi He

"Nei" means *internal*. "Nei San Bao" means *three internal treasures*, which are Shen, Qi, and Jing. "Zi He" means *join together*. "Nei San Bao Zi He" means *three internal treasures (Shen, Qi, Jing) join as one*. When the three internal treasures join as one, a Jin Dan is formed. It is Ming Xin Jian Xing, enlighten your heart to see your own true nature. During this process, your Shi Shen (reincarnated soul) will realize and receive guidance from your Yuan Shen (original Tao soul). Gradually, Shi Shen will align more and more with Yuan Shen until, finally, they meld together as one. This is a very long process, but an essential step to reach Tao.

外不漏 Wai Bu Lou
Three external treasures (eyes, ears, mouth) do not leak.

內自合 Nei Zi He
Three internal treasures (Shen, Qi, Jing) join as one.

通天達地 Tong Tian Da Di
"Tong Tian" means *connect with Heaven*. "Da Di" means *connect with Mother Earth*. "Tong Tian Da Di" means *connect through Heaven and Mother Earth as one*. When you join your three internal treasures (Shen, Qi, Jing) together, you will see your Yuan Shen, your original soul, which is Tao. Tao creates Heaven, Mother Earth, and all things, which means you have connected with them. They are in you. You are in them. Everyone and everything carries the Tao nature. Why can't a person reach Tao? It is because of the person's Shen Qi Jing blockages.

逍遙道中 Xiao Yao Dao Zhong
"Xiao Yao" means *free, liberated*. "Dao Zhong" means *in the Tao*. "Xiao Yao Dao Zhong" means *meld with Tao to have true freedom from all You World laws*. After your three external treasures (eyes, ears, mouth) stop leaking and your three internal treasures (Shen, Qi, Jing) join as one, you have reached the level where Heaven and Mother Earth join as one with you. Then you can be uplifted from the You (existence) World to the Wu (emptiness) World. Therefore, you are no longer controlled by the laws of Heaven and Mother Earth. You have reached the Tao. You have reached the condition that Lao Zi expressed as Dao Fa Zi Ran, *follow nature's way*. You will be able to offer great service with all Ten Da Tao Natures.

Practice

Apply the Four Power Techniques:

1. Body Power

Sit or stand as in the practice in chapter one. Alternate sitting and standing positions in your practice for best results.

2. Soul Power

Say *hello* to inner and outer souls:
> *Dear Tao Source,*
> *Dear Heaven and Mother Earth,*
> *Dear countless planets, stars, galaxies, and universes,*
> *Dear Bao Yuan Shou Yi Chang Shou Yong Sheng Dao Jing* (Tao Classic of
>   Longevity and Immortality),
> *I love you, honor you, and appreciate you.*
> *Please purify and remove my Shen Qi Jing blockages for healing, rejuvenation,*
>   *transformation of relationships and finances, increasing intelligence, and*
>   *flourishing in every aspect of my life.*
> *Please bless my spiritual practice of longevity and immortality, close my*
>   *external treasures (eyes, ears, mouth), and join my three internal treasures*
>   *(Shen Qi Jing) as one.*
> *Please bless me to achieve my longevity and immortality journey.*
> *I am extremely honored and grateful.*
> *Thank you. Thank you. Thank you.*

3. Mind Power

Visualize a Tao Source Jin Dan (golden light ball) shining and rotating in the Lower Dan Tian area (figure 16). The light is cleansing Shen Qi Jing blockages in your Lower Dan Tian area and then getting stronger and brighter as the Jin Dan grows to the size of the body.

Figure 16. Wai Bu Lou, Nei Zi He Jin Dan Practice

4. Sound Power

Chant silently or aloud the first thirteen stanzas of the *Tao Classic of Longevity and Immortality* (from chapters one through thirteen), and then repeat the thirteenth stanza (in **bold** below) ten or more times in each practice session. Alternate silent and aloud (Yin Yang) chanting for best results.

Hei Hei Hei
Hei Jin Dan
Da Guang Ming
Da Guang Ming Jin Dan
Wo Zai Dao Guang Zhong
Dao Guang Zai Wo Zhong
Tong Ti Tou Ming

Heng Heng Heng
Heng Jin Dan
Da Gan En

Da Gan En Jin Dan
Dao Sheng De Yang
Zai Pei Ci Hui
Dao En Yong Cun

Hong Hong Hong
Hong Jin Dan
Da Qian Bei
Da Qian Bei Jin Dan
Rou Ruo Bu Zheng
Chi Xu Jing Jin
Shi Qian Bei
Die Wan Zhang

Ah Ah Ah
Ah Jin Dan
Da Ai
Da Ai Jin Dan
Wu Tiao Jian Ai
Rong Hua Zai Nan
Xin Qing Shen Ming

Xi Xi Xi
Xi Jin Dan
Da Fu Wu
Da Fu Wu Jin Dan
Shi Wei Gong Pu
Wu Si Feng Xian
Shang Cheng Fa Men

Yi Yi Yi
Yi Jin Dan
Da Kuan Shu
Da Kuan Shu Jin Dan
Wo Yuan Liang Ni
Ni Yuan Liang Wo
Xiang Ai Ping An He Xie

Weng Weng Weng
Weng Jin Dan
Da Yuan Man
Da Yuan Man Jin Dan
Ling Xin Nao Shen Yuan Man
Ren Di Tian Dao Shen Xian Ti
Fu Wu Xiu Lian Cai Ke Pan

You You You
You Jin Dan
Da Ci Bei
Da Ci Bei Jin Dan
Yuan Li Zeng Qiang
Fu Wu Zhong Sheng
Gong De Wu Liang

Ha Ha Ha
Ha Jin Dan
Da Chang Sheng
Da Chang Sheng Jin Dan
Dao Ci Ying Fu
Xing Shan Ji De
Dao Ye Chang Sheng

Yu Yu Yu
Yu Jin Dan
Da He Xie
Da He Xie Jin Dan
San Ren Tong Xin
Qi Li Duan Jin
Cheng Gong Mi Jue

Hei Heng Hong Ah Xi Yi Weng You
Hei Heng Hong Ah Xi Yi Weng You
Hei Heng Hong Ah Xi Yi Weng You
Hei Heng Hong Ah Xi Yi Weng You

You Weng Yi Xi Ah Hong Heng Hei
You Weng Yi Xi Ah Hong Heng Hei
You Weng Yi Xi Ah Hong Heng Hei
You Weng Yi Xi Ah Hong Heng Hei

Weng Hei Hong You
Weng Hei Hong You
Weng Hei Hong You
Weng Hei Hong You

Guang Gan Qian Ai Fu Kuan Yuan Ci
Guang Gan Qian Ai Fu Kuan Yuan Ci
Guang Gan Qian Ai Fu Kuan Yuan Ci
Guang Gan Qian Ai Fu Kuan Yuan Ci

Ci Yuan Kuan Fu Ai Qian Gan Guang
Ci Yuan Kuan Fu Ai Qian Gan Guang
Ci Yuan Kuan Fu Ai Qian Gan Guang
Ci Yuan Kuan Fu Ai Qian Gan Guang

Yuan Guang Qian Ci
Yuan Guang Qian Ci
Yuan Guang Qian Ci
Yuan Guang Qian Ci

Xu Xu Xu
Xu Jin Dan
Shu Gan Jin Dan
Ah Ah Ah
Ah Jin Dan
Yang Xin Jin Dan

Hu Hu Hu
Hu Jin Dan
Jian Pi Jin Dan

Si Si Si
Si Jin Dan
Xuan Fei Jin Dan

Chui Chui Chui
Chui Jin Dan
Zhuang Shen Jin Dan (1 time)

Mu Bu Wang Shi
Er Bu Wang Ting
Kou Bu Wang Yan
Wai San Bao Bu Lou
Bu Shi An Shen Yu Xin
Bu Ting Xu Jing Yu Shen
Bu Yan Yun Qi Dan Tian
Nei San Bao Zi He
Wai Bu Lou
Nei Zi He
Tong Tian Da Di
Xiao Yao Dao Zhong (10 times)

Part III
Attain Tao

BAO YUAN SHOU Yi is the final step to reach Tao and become a Buddha. This is the highest pursuit and highest goal in one's physical life and spiritual life.

Remember the profound sacred wisdom of Tao Normal Creation and Tao Reverse Creation that I shared in the introduction to this book and in figure 1.

Tao and One are the Wu World (emptiness and formless world). In the Wu World, the law is Bao Yuan Shou Yi, Oneness. Heaven and Mother Earth belong to the You World (existence and forms world). The number one law in the You World is Yin Yang law. We live on Mother Earth, so we live in the Yin Yang world. Hence, birth, aging, sickness, and death are the inescapable parts of life. To escape from those sufferings and become an immortal, one must apply the immortal law, which is the Wu World law of Bao Yuan Shou Yi, to become Oneness.

In part III, chapters fourteen through nineteen, I will teach you the sacred wisdom of Bao Yuan Shou Yi, and how to practice it to form and grow the Jin Dan in your body to achieve longevity and immortality, which is to reach Tao, the summit of all sacred spiritual mountains.

Chapter Fourteen

Chinese	Pinyin	English Translation
抱元守一	Bao Yuan Shou Yi	*Hold Tao, focus on One, which is Tao*
一即是中	Yi Ji Shi Zhong	*One is Zhong, the core of one's life, which is Tao*
坤宮命門尾閭 會陰足跟中	Kun Gong Ming Men Wei Lü Hui Yin Zu Gen Zhong	*Kun Temple, Ming Men, sacrum, perineum, and heels of feet Zhong*
人中地中	Ren Zhong Di Zhong	*Zhong includes Human Zhong, Mother Earth Zhong*
天中道中	Tian Zhong Dao Zhong	*Zhong also includes Heaven Zhong, Tao Zhong*
中中道空空	Zhong Zhong Dao Kong Kong	*Core, core, Tao, emptiness, emptiness, which are all Tao Source*

Sacred wisdom

抱元守一 Bao Yuan Shou Yi
I have explained this phrase in the introduction. "Bao Yuan Shou Yi" means *to hold Yuan Shen, Yuan Qi, and Yuan Jing and focus on One, which is Tao.*

一即是中 Yi Ji Shi Zhong
"Yi" means *Oneness*. "Ji Shi" means *is exactly*. "Zhong" means *core of life*. "Yi Ji Shi Zhong" means *Oneness is Zhong, the core of life, which is Tao.*

坤宮命門尾閭會陰足根中 Kun Gong Ming Men Wei Lü Hui Yin Zu Gen
Zhong

This sacred mantra describes the position of the Zhong (figure 17) in a human
body.

"Kun Gong" means *Kun Temple*. It is located in the third Soul House, in the
center of the body behind the navel.

"Ming Men" refers to the Ming Men acupuncture point on the back. It is
directly behind the navel. "Ming" means *life*. "Men" means *gate*. "Ming Men"
means *life gate*. It is the core of the Kundalini, a foundational energy center that
is located in front of the sacrum.

"Wei Lü" is the sacrum or tailbone area.

"Hui Yin" is an acupuncture point located on the perineum, between the
genitals and the anus.

"Zu Gen" means *the heels of both feet*.

"Kun Gong Ming Men Wei Lü Hui Yin Zu Gen Zhong" means *the Zhong
extends from the Kun Temple to the Ming Men acupuncture point, to the tailbone, to
the first Soul House, and down to the heels of both feet*.

人中地中 Ren Zhong Di Zhong

"Ren" means *human being*. "Zhong" means *core of life*. "Di" means *Mother Earth*.
"Ren Zhong Di Zhong" means *a human being's Zhong is located here. Mother
Earth's Zhong is also located here*.

天中道中 Tian Zhong Dao Zhong

"Tian" means *Heaven*. Dao is the Source or Ultimate Creator. "Tian Zhong Dao
Zhong" means that *this Zhong is also the Zhong of Heaven and Tao*.

中中道空空 Zhong Zhong Dao Kong Kong

"Kong" means *emptiness*. "Zhong Zhong Dao Kong Kong" means *core, core, Tao,
emptiness, emptiness, which are all Tao Source*.

Practice

Apply the Four Power Techniques:

1. Body Power

Sit or stand as in the practice in chapter one. Alternate sitting and standing positions in your practice for the best results.

2. Soul Power

Say *hello* to inner and outer souls:

> *Dear Tao Source,*
> *Dear Heaven and Mother Earth,*
> *Dear countless planets, stars, galaxies, and universes,*
> *Dear Bao Yuan Shou Yi Chang Shou Yong Sheng Dao Jing* (Tao Classic of
>   Longevity and Immortality),
> *I love you, honor you, and appreciate you.*
> *Please purify and remove my Shen Qi Jing blockages for healing, rejuvenation,*
>   *transformation of relationships and finances, increasing intelligence, and*
>   *flourishing in every aspect of my life.*
> *Please bless my spiritual practice of longevity and immortality and develop*
>   *my Zhong.*
> *Please bless me to achieve longevity and immortality.*
> *I am extremely honored and grateful.*
> *Thank you. Thank you. Thank you.*

3. Mind Power

Visualize a Tao Source Jin Dan (golden light ball) shining and rotating in your Zhong area. The light is cleansing Shen Qi Jing blockages in your Zhong area and getting brighter and brighter as it grows to the size of the body (figure 17).

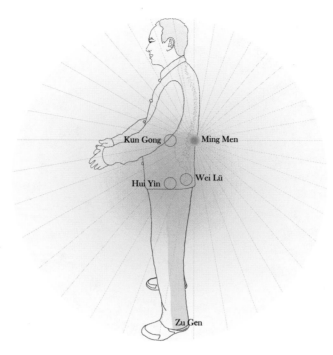

Figure 17. Zhong Practice

4. Sound Power

Chant silently or aloud the first fourteen stanzas of the *Tao Classic of Longevity and Immortality* (from chapters one through fourteen), and then repeat the fourteenth stanza (in **bold** below) ten or more times in each practice session. Alternate silent and aloud (Yin Yang) chanting for best results.

Hei Hei Hei
Hei Jin Dan
Da Guang Ming
Da Guang Ming Jin Dan
Wo Zai Dao Guang Zhong
Dao Guang Zai Wo Zhong
Tong Ti Tou Ming

Heng Heng Heng
Heng Jin Dan
Da Gan En
Da Gan En Jin Dan
Dao Sheng De Yang
Zai Pei Ci Hui
Dao En Yong Cun

Hong Hong Hong
Hong Jin Dan
Da Qian Bei
Da Qian Bei Jin Dan
Rou Ruo Bu Zheng
Chi Xu Jing Jin
Shi Qian Bei
Die Wan Zhang

Ah Ah Ah
Ah Jin Dan
Da Ai
Da Ai Jin Dan
Wu Tiao Jian Ai
Rong Hua Zai Nan
Xin Qing Shen Ming

Xi Xi Xi
Xi Jin Dan
Da Fu Wu
Da Fu Wu Jin Dan
Shi Wei Gong Pu
Wu Si Feng Xian
Shang Cheng Fa Men

Yi Yi Yi
Yi Jin Dan
Da Kuan Shu
Da Kuan Shu Jin Dan

Wo Yuan Liang Ni
Ni Yuan Liang Wo
Xiang Ai Ping An He Xie

Weng Weng Weng
Weng Jin Dan
Da Yuan Man
Da Yuan Man Jin Dan
Ling Xin Nao Shen Yuan Man
Ren Di Tian Dao Shen Xian Ti
Fu Wu Xiu Lian Cai Ke Pan

You You You
You Jin Dan
Da Ci Bei
Da Ci Bei Jin Dan
Yuan Li Zeng Qiang
Fu Wu Zhong Sheng
Gong De Wu Liang

Ha Ha Ha
Ha Jin Dan
Da Chang Sheng
Da Chang Sheng Jin Dan
Dao Ci Ying Fu
Xing Shan Ji De
Dao Ye Chang Sheng

Yu Yu Yu
Yu Jin Dan
Da He Xie
Da He Xie Jin Dan
San Ren Tong Xin
Qi Li Duan Jin
Cheng Gong Mi Jue

Hei Heng Hong Ah Xi Yi Weng You
Hei Heng Hong Ah Xi Yi Weng You
Hei Heng Hong Ah Xi Yi Weng You
Hei Heng Hong Ah Xi Yi Weng You

You Weng Yi Xi Ah Hong Heng Hei
You Weng Yi Xi Ah Hong Heng Hei
You Weng Yi Xi Ah Hong Heng Hei
You Weng Yi Xi Ah Hong Heng Hei

Weng Hei Hong You
Weng Hei Hong You
Weng Hei Hong You
Weng Hei Hong You

Guang Gan Qian Ai Fu Kuan Yuan Ci
Guang Gan Qian Ai Fu Kuan Yuan Ci
Guang Gan Qian Ai Fu Kuan Yuan Ci
Guang Gan Qian Ai Fu Kuan Yuan Ci

Ci Yuan Kuan Fu Ai Qian Gan Guang
Ci Yuan Kuan Fu Ai Qian Gan Guang
Ci Yuan Kuan Fu Ai Qian Gan Guang
Ci Yuan Kuan Fu Ai Qian Gan Guang

Yuan Guang Qian Ci
Yuan Guang Qian Ci
Yuan Guang Qian Ci
Yuan Guang Qian Ci

Xu Xu Xu
Xu Jin Dan
Shu Gan Jin Dan

Ah Ah Ah
Ah Jin Dan
Yang Xin Jin Dan

Hu Hu Hu
Hu Jin Dan
Jian Pi Jin Dan

Si Si Si
Si Jin Dan
Xuan Fei Jin Dan

Chui Chui Chui
Chui Jin Dan
Zhuang Shen Jin Dan

Mu Bu Wang Shi
Er Bu Wang Ting
Kou Bu Wang Yan
Wai San Bao Bu Lou
Bu Shi An Shen Yu Xin
Bu Ting Xu Jing Yu Shen
Bu Yan Yun Qi Dan Tian
Nei San Bao Zi He
Wai Bu Lou
Nei Zi He
Tong Tian Da Di
Xiao Yao Dao Zhong (1 time)

Bao Yuan Shou Yi
Yi Ji Shi Zhong
Kun Gong Ming Men Wei Lü Hui Yin Zu Gen Zhong
Ren Zhong Di Zhong
Tian Zhong Dao Zhong
Zhong Zhong Dao Kong Kong (10 times)

Chapter Fifteen

Chinese	Pinyin	English Translation
抱元守一	Bao Yuan Shou Yi	Hold Tao, focus on One, which is Tao
天一真水	Tian Yi Zhen Shui	Heaven's Oneness sacred nectar
金津玉液	Jin Jin Yu Ye	Earth's sacred golden and jade nectar
咽入丹田	Yan Ru Dan Tian	Swallow Heaven's and Earth's sacred nectars into the Lower Dan Tian
神氣精合一	Shen Qi Jing He Yi	Soul, heart, mind, energy, and matter join as one
天地人合一	Tian Di Ren He Yi	Heaven's, Mother Earth's, human's soul, heart, mind, energy, and matter join as one
金丹形成	Jin Dan Xing Cheng	Jin Dan is formed

Sacred wisdom

抱元守一 Bao Yuan Shou Yi

Yuan includes Yuan Shen, Yuan Qi, and Yuan Jing. "Bao Yuan Shou Yi" means *hold your Yuan Shen, Yuan Qi, and Yuan Jing, which are Tao. Focus on Oneness, which is also Tao.* In fact, Bao Yuan Shou Yi is to hold Tao and focus on One.

天一真水 Tian Yi Zhen Shui

"Tian" means *Heaven*. "Yi" means *Oneness*. "Zhen" means *sacred*. "Shui" means *liquid* or *water*. "Tian Yi Zhen Shui" means *Heaven's Oneness sacred liquid*. This liquid falls from Heaven into the seventh Soul House (Bai Hui acupuncture point at the crown), and then goes into the brain and mouth.

金津玉液 Jin Jin Yu Ye

The first "Jin" in this phrase means *golden*. The second "Jin" is a different Chinese character that means *liquid*. "Yu" means *jade*. "Ye" is another character for *liquid*. "Jin Jin Yu Ye" means *golden and jade liquid*, which implies that the liquid is invaluable. This sacred liquid comes from Mother Earth. It goes up through the Yong Quan acupuncture points on the soles of the feet, the legs, the central meridian, and finally up to the salivary glands.

咽入丹田 Yan Ru Dan Tian

"Yan" means *swallow*. "Ru" means *to go into*. "Dan Tian" refers to the Lower Dan Tian. "Yan Ru Dan Tian" means *swallow Tian Yi Zhen Shui (Heaven's Oneness sacred liquid) and Jin Jin Yu Ye (Mother Earth's sacred liquid) into the Lower Dan Tian*.

神氣精合一 Shen Qi Jing He Yi

Shen includes soul, heart, and mind. "Qi" means *energy*. "Jing" means *matter*. "He" means *join as*. "Yi" means *one*. "Shen Qi Jing He Yi" means *soul, heart, mind, energy, and matter join as one*.

天地人合一 Tian Di Ren He Yi

"Tian" means *Heaven*. "Di" means *Mother Earth*. "Ren" means *human being*. "He" means *join as*. "Yi" means *one*. "Tian Di Ren He Yi" means *Heaven's, Mother Earth's, and human being's Shen Qi Jing join as one*.

金丹形成 Jin Dan Xing Cheng

"Jin Dan" means *golden light ball*. "Xing Cheng" means *formed*. A normal human being does not have a Jin Dan. Only a Tao practitioner can form a Jin Dan in the body through sacred spiritual practice. If you have followed my teachings in this book and have been practicing seriously, you may very well have already formed a Jin Dan in your lower abdomen.

Practice

Apply the Four Power Techniques:

1. Body Power

Sit or stand as in the practice in chapter one. Alternate sitting and standing positions in your practice for the best results.

2. Soul Power

Say *hello* to inner and outer souls:

> *Dear Tao Source,*
> *Dear Heaven and Mother Earth,*
> *Dear countless planets, stars, galaxies, and universes,*
> *Dear Bao Yuan Shou Yi Chang Shou Yong Sheng Dao Jing* (Tao Classic of
>   Longevity and Immortality),
> *I love you, honor you, and appreciate you.*
> *Please purify and remove my Shen Qi Jing blockages for healing, rejuvenation,*
>   *transformation of relationships and finances, increasing intelligence, and*
>   *flourishing in every aspect of my life.*
> *Please bless my spiritual practice of longevity and immortality, and form and*
>   *develop a Jin Dan in my body.*
> *Please bless me to achieve longevity and immortality.*
> *I am extremely honored and grateful.*
> *Thank you. Thank you. Thank you.*

3. Mind Power

Visualize a Tao Source Jin Dan (golden light ball) shining and rotating in your Zhong. The light is cleansing Shen Qi Jing blockages in your Zhong and then getting brighter and brighter as it grows to the size of your body (figure 18).

Figure 18. Form and Develop Jin Dan Practice

4. *Sound Power*

Chant silently or aloud the first fifteen stanzas of the *Tao Classic of Longevity and Immortality* (from chapter one through chapter fifteen), and then repeat the fifteenth stanza (in **bold** below) ten or more times in each practice session. Alternate silent and aloud (Yin Yang) chanting for best results.

Hei Hei Hei
Hei Jin Dan
Da Guang Ming
Da Guang Ming Jin Dan
Wo Zai Dao Guang Zhong
Dao Guang Zai Wo Zhong
Tong Ti Tou Ming

Heng Heng Heng
Heng Jin Dan
Da Gan En

Da Gan En Jin Dan
Dao Sheng De Yang
Zai Pei Ci Hui
Dao En Yong Cun

Hong Hong Hong
Hong Jin Dan
Da Qian Bei
Da Qian Bei Jin Dan
Rou Ruo Bu Zheng
Chi Xu Jing Jin
Shi Qian Bei
Die Wan Zhang

Ah Ah Ah
Ah Jin Dan
Da Ai
Da Ai Jin Dan
Wu Tiao Jian Ai
Rong Hua Zai Nan
Xin Qing Shen Ming

Xi Xi Xi
Xi Jin Dan
Da Fu Wu
Da Fu Wu Jin Dan
Shi Wei Gong Pu
Wu Si Feng Xian
Shang Cheng Fa Men

Yi Yi Yi
Yi Jin Dan
Da Kuan Shu
Da Kuan Shu Jin Dan
Wo Yuan Liang Ni
Ni Yuan Liang Wo
Xiang Ai Ping An He Xie

Weng Weng Weng
Weng Jin Dan
Da Yuan Man
Da Yuan Man Jin Dan
Ling Xin Nao Shen Yuan Man
Ren Di Tian Dao Shen Xian Ti
Fu Wu Xiu Lian Cai Ke Pan

You You You
You Jin Dan
Da Ci Bei
Da Ci Bei Jin Dan
Yuan Li Zeng Qiang
Fu Wu Zhong Sheng
Gong De Wu Liang

Ha Ha Ha
Ha Jin Dan
Da Chang Sheng
Da Chang Sheng Jin Dan
Dao Ci Ying Fu
Xing Shan Ji De
Dao Ye Chang Sheng

Yu Yu Yu
Yu Jin Dan
Da He Xie
Da He Xie Jin Dan
San Ren Tong Xin
Qi Li Duan Jin
Cheng Gong Mi Jue

Hei Heng Hong Ah Xi Yi Weng You
Hei Heng Hong Ah Xi Yi Weng You
Hei Heng Hong Ah Xi Yi Weng You
Hei Heng Hong Ah Xi Yi Weng You

You Weng Yi Xi Ah Hong Heng Hei
You Weng Yi Xi Ah Hong Heng Hei
You Weng Yi Xi Ah Hong Heng Hei
You Weng Yi Xi Ah Hong Heng Hei

Weng Hei Hong You
Weng Hei Hong You
Weng Hei Hong You
Weng Hei Hong You

Guang Gan Qian Ai Fu Kuan Yuan Ci
Guang Gan Qian Ai Fu Kuan Yuan Ci
Guang Gan Qian Ai Fu Kuan Yuan Ci
Guang Gan Qian Ai Fu Kuan Yuan Ci

Ci Yuan Kuan Fu Ai Qian Gan Guang
Ci Yuan Kuan Fu Ai Qian Gan Guang
Ci Yuan Kuan Fu Ai Qian Gan Guang
Ci Yuan Kuan Fu Ai Qian Gan Guang

Yuan Guang Qian Ci
Yuan Guang Qian Ci
Yuan Guang Qian Ci
Yuan Guang Qian Ci

Xu Xu Xu
Xu Jin Dan
Shu Gan Jin Dan

Ah Ah Ah
Ah Jin Dan
Yang Xin Jin Dan

Hu Hu Hu
Hu Jin Dan
Jian Pi Jin Dan

Si Si Si
Si Jin Dan
Xuan Fei Jin Dan

Chui Chui Chui
Chui Jin Dan
Zhuang Shen Jin Dan

Mu Bu Wang Shi
Er Bu Wang Ting
Kou Bu Wang Yan
Wai San Bao Bu Lou
Bu Shi An Shen Yu Xin
Bu Ting Xu Jing Yu Shen
Bu Yan Yun Qi Dan Tian
Nei San Bao Zi He
Wai Bu Lou
Nei Zi He
Tong Tian Da Di
Xiao Yao Dao Zhong

Bao Yuan Shou Yi
Yi Ji Shi Zhong
Kun Gong Ming Men Wei Lü Hui Yin Zu Gen Zhong
Ren Zhong Di Zhong
Tian Zhong Dao Zhong
Zhong Zhong Dao Kong Kong (1 time)

Bao Yuan Shou Yi
Tian Yi Zhen Shui
Jin Jin Yu Ye
Yan Ru Dan Tian
Shen Qi Jing He Yi
Tian Di Ren He Yi
Jin Dan Xing Cheng (10 times)

Chapter Sixteen

Chinese	Pinyin	English Translation
抱元守一	Bao Yuan Shou Yi	*Hold Tao, focus on One, which is Tao*
舌抵上顎	She Di Shang E	*Tip of tongue touches palate*
咽津不斷	Yan Jin Bu Duan	*Swallow Heaven's and Earth's sacred nectars constantly*
金丹壯大	Jin Dan Zhuang Da	*Jin Dan grows bigger and stronger*

Sacred wisdom

抱元守一 Bao Yuan Shou Yi
Hold Tao, focus on One.

舌抵上顎 She Di Shang E
"She" means *tongue*. "Di" means *touch*. "Shang E" means *palate*. "She Di Shang E" means *hold the tip of your tongue to the roof of your mouth*. The sacred wisdom within this phrase and practice is that it connects two major meridians, the Ren meridian (任脈) and the Du meridian (督脈). The Ren meridian, or Conception Vessel, runs up the center line of the front of the body. It is the conception vessel of all the major Yin meridians, including the liver, heart, spleen, lung, kidney, and pericardium meridians. The Du meridian, or Governing Vessel, runs along the center of the back, following the path of the spinal cord up from the tailbone area to the top of the head, and then ends inside the mouth. The Du meridian governs the major Yang meridians, including the gallbladder, small intestine, stomach, large

intestine, urinary bladder, and San Jiao meridians. The San Jiao is the pathway of Qi and bodily fluid. It is divided into Lower Jiao (space from pelvic floor up to the level of the navel), Middle Jiao (space from level of navel up to the diaphragm), and Upper Jiao (space above the diaphragm). When the tip of the tongue touches the palate, it connects the Ren meridian and Du meridian, promoting the flow of energy and bodily fluid through the Ren Du meridians. To promote Qi flow and blood flow in the Yin and Yang meridians is to balance Yin and Yang, which is the most important five-thousand-year-old principle of healing in traditional Chinese medicine. This is a major sacred practice for the Tao journey.

咽津不斷 Yan Jin Bu Duan
"Yan" means *swallow*. "Jin" includes Tian Yi Zhen Shui and Jin Jin Yu Ye, Heaven's and Mother Earth's sacred liquids, respectively. "Bu Duan" means *nonstop*. "Yan Jin Bu Duan" means *swallow Tian Yi Zhen Shui and Jin Jin Yu Ye nonstop*. These liquids are not the normal saliva of the human body. You can only receive Heaven's sacred liquid (Tian Yi Zhen Shui) and Mother Earth's sacred liquid (Jin Jin Yu Ye) through sacred spiritual practices. They are the sacred essential ingredients to form and grow the Jin Dan. Jin Dan is a golden light ball. Growing this Jin Dan *is* the Tao journey. The size of the Jin Dan represents the achievements of one's Tao journey.

金丹壯大 Jin Dan Zhuang Da
"Zhuang Da" means *to expand*. "Jin Dan Zhuang Da" means *the golden light ball expands and becomes stronger*.

Practice

Apply the Four Power Techniques:

1. Body Power

Sit or stand as in the practice in chapter one. Alternate sitting and standing positions in your practice for the best results.

2. Soul Power

Say *hello* to inner and outer souls:

> *Dear Tao Source,*
> *Dear Heaven and Mother Earth,*
> *Dear countless planets, stars, galaxies, and universes,*
> *Dear Bao Yuan Shou Yi Chang Shou Yong Sheng Dao Jing* (Tao Classic of
>   Longevity and Immortality),
> *I love you, honor you, and appreciate you.*
> *Please purify and remove my Shen Qi Jing blockages for healing, rejuvenation,*
>   *transformation of relationships and finances, increasing intelligence, and*
>   *flourishing in every aspect of my life.*
> *Please bless my spiritual practice of longevity and immortality and grow and*
>   *expand my Jin Dan.*
> *Please bless me to achieve my longevity and immortality journey.*
> *I am extremely honored and grateful.*
> *Thank you. Thank you. Thank you.*

3. Mind Power

Visualize a Tao Source Jin Dan (golden light ball) shining and rotating in your Zhong. The light is cleansing Shen Qi Jing blockages in your Zhong and then getting brighter and brighter as the Jin Dan grows to the size of the body (figure 19).

4. Sound Power

Chant silently or aloud the first sixteen stanzas of the *Tao Classic of Longevity and Immortality* (from chapters one through sixteen), and then repeat the sixteenth stanza (in **bold** below) ten or more times in each practice session. Alternate silent and aloud (Yin Yang) chanting for optimum benefits.

> *Hei Hei Hei*
> *Hei Jin Dan*

Figure 19. Develop and Grow Jin Dan Practice

Da Guang Ming
Da Guang Ming Jin Dan
Wo Zai Dao Guang Zhong
Dao Guang Zai Wo Zhong
Tong Ti Tou Ming

Heng Heng Heng
Heng Jin Dan
Da Gan En
Da Gan En Jin Dan
Dao Sheng De Yang
Zai Pei Ci Hui
Dao En Yong Cun

Hong Hong Hong
Hong Jin Dan
Da Qian Bei
Da Qian Bei Jin Dan
Rou Ruo Bu Zheng

Chi Xu Jing Jin
Shi Qian Bei
Die Wan Zhang

Ah Ah Ah
Ah Jin Dan
Da Ai
Da Ai Jin Dan
Wu Tiao Jian Ai
Rong Hua Zai Nan
Xin Qing Shen Ming

Xi Xi Xi
Xi Jin Dan
Da Fu Wu
Da Fu Wu Jin Dan
Shi Wei Gong Pu
Wu Si Feng Xian
Shang Cheng Fa Men

Yi Yi Yi
Yi Jin Dan
Da Kuan Shu
Da Kuan Shu Jin Dan
Wo Yuan Liang Ni
Ni Yuan Liang Wo
Xiang Ai Ping An He Xie

Weng Weng Weng
Weng Jin Dan
Da Yuan Man
Da Yuan Man Jin Dan
Ling Xin Nao Shen Yuan Man
Ren Di Tian Dao Shen Xian Ti
Fu Wu Xiu Lian Cai Ke Pan

You You You

You Jin Dan
Da Ci Bei
Da Ci Bei Jin Dan
Yuan Li Zeng Qiang
Fu Wu Zhong Sheng
Gong De Wu Liang

Ha Ha Ha
Ha Jin Dan
Da Chang Sheng
Da Chang Sheng Jin Dan
Dao Ci Ying Fu
Xing Shan Ji De
Dao Ye Chang Sheng

Yu Yu Yu
Yu Jin Dan
Da He Xie
Da He Xie Jin Dan
San Ren Tong Xin
Qi Li Duan Jin
Cheng Gong Mi Jue

Hei Heng Hong Ah Xi Yi Weng You
Hei Heng Hong Ah Xi Yi Weng You
Hei Heng Hong Ah Xi Yi Weng You
Hei Heng Hong Ah Xi Yi Weng You

You Weng Yi Xi Ah Hong Heng Hei
You Weng Yi Xi Ah Hong Heng Hei
You Weng Yi Xi Ah Hong Heng Hei
You Weng Yi Xi Ah Hong Heng Hei

Weng Hei Hong You
Weng Hei Hong You
Weng Hei Hong You
Weng Hei Hong You

Guang Gan Qian Ai Fu Kuan Yuan Ci
Guang Gan Qian Ai Fu Kuan Yuan Ci
Guang Gan Qian Ai Fu Kuan Yuan Ci
Guang Gan Qian Ai Fu Kuan Yuan Ci

Ci Yuan Kuan Fu Ai Qian Gan Guang
Ci Yuan Kuan Fu Ai Qian Gan Guang
Ci Yuan Kuan Fu Ai Qian Gan Guang
Ci Yuan Kuan Fu Ai Qian Gan Guang

Yuan Guang Qian Ci
Yuan Guang Qian Ci
Yuan Guang Qian Ci
Yuan Guang Qian Ci

Xu Xu Xu
Xu Jin Dan
Shu Gan Jin Dan

Ah Ah Ah
Ah Jin Dan
Yang Xin Jin Dan

Hu Hu Hu
Hu Jin Dan
Jian Pi Jin Dan

Si Si Si
Si Jin Dan
Xuan Fei Jin Dan

Chui Chui Chui
Chui Jin Dan
Zhuang Shen Jin Dan

Mu Bu Wang Shi
Er Bu Wang Ting

Kou Bu Wang Yan
Wai San Bao Bu Lou
Bu Shi An Shen Yu Xin
Bu Ting Xu Jing Yu Shen
Bu Yan Yun Qi Dan Tian
Nei San Bao Zi He
Wai Bu Lou
Nei Zi He
Tong Tian Da Di
Xiao Yao Dao Zhong

Bao Yuan Shou Yi
Yi Ji Shi Zhong
Kun Gong Ming Men Wei Lü Hui Yin Zu Gen Zhong
Ren Zhong Di Zhong
Tian Zhong Dao Zhong
Zhong Zhong Dao Kong Kong

Bao Yuan Shou Yi
Tian Yi Zhen Shui
Jin Jin Yu Ye
Yan Ru Dan Tian
Shen Qi Jing He Yi
Tian Di Ren He Yi
Jin Dan Xing Cheng (1 time)

Bao Yuan Shou Yi
She Di Shang E
Yan Jin Bu Duan
Jin Dan Zhuang Da (10 times)

Chapter Seventeen

Chinese	Pinyin	English Translation
抱元守一	Bao Yuan Shou Yi	*Hold Tao, focus on One, which is Tao*
十圓滿合一	Shi Yuan Man He Yi	*Number ten indicates soul, heart, mind, and body enlightenments join as one*
九九歸一	Jiu Jiu Gui Yi	*Number nine indicates all kinds of spiritual belief systems join as one*
八卦合一	Ba Gua He Yi	*Number eight indicates I Ching Ba Gua, including Heaven, Earth, thunder, wind, water, fire, mountain, and lake, join as one*
北斗七星合一	Bei Dou Qi Xing He Yi	*Number seven indicates the seven stars in the Big Dipper join as one*
南斗六星合一	Nan Dou Liu Xing He Yi	*Number six indicates the six stars in the Southern Dipper join as one*
五行合一	Wu Xing He Yi	*Number five indicates Five Elements join as one*

四象合一	Si Xiang He Yi	*Number four indicates different conditions of Yin and Yang, including Young Yin, Old Yin, Young Yang, and Old Yang, join as one*
神氣精合一	Shen Qi Jing He Yi	*Soul, heart, mind, energy, matter join as one*
陰陽合一	Yin Yang He Yi	*Yin Yang join as one*
人地天道中合一	Ren Di Tian Dao Zhong He Yi	*Zhong of human being, Mother Earth, Heaven, and Tao join as one*
萬物合一	Wan Wu He Yi	*Countless planets, stars, galaxies, and universes, human beings, and all things join as one*

Sacred wisdom

抱元守一 Bao Yuan Shou Yi
Hold Tao (your Yuan Shen, Yuan Qi, and Yuan Jing) as One and focus on One.

十圓滿合一 Shi Yuan Man He Yi
"Shi" means *ten*. The message of ten is "completeness." Yuan Man includes soul, heart, mind, and body enlightenment, which is the highest enlightenment and the highest achievement in one's Tao journey. "He Yi" means *join as one*. "Shi Yuan Man He Yi" means *complete enlightenment of soul, heart, mind, and body with Tao*.

九九歸一 Jiu Jiu Gui Yi
"Jiu" means *nine*. "Gui" means *to return*. "Yi" means *Oneness*. "Jiu Jiu Gui Yi" indicates all kinds of spiritual belief systems join as one.

八卦合一 Ba Gua He Yi
"Ba Gua" is presented in *I Ching*, the secret ancient wisdom and practice. Ba Gua represents eight natures: Heaven, Earth, Thunder, Wind, Water, Fire, Mountain, and Lake (figure 20). Each trigram carries unique Shen Qi Jing. "Ba

Gua He Yi" means *the Shen Qi Jing of all eight natures join as one*. This phrase carries phenomenal power for reaching Tao.

Figure 20. Ba Gua

北斗七星合一 Bei Dou Qi Xing He Yi

"Bei Dou" means *the Big Dipper* (figure 21). "Qi" means *seven*. "Xing" means *star*. "Bei Dou Qi Xing He Yi" means *the Shen Qi Jing of the seven stars in the Big Dipper join as one*. The Big Dipper consists of seven bright stars. In sacred spiritual wisdom, the Big Dipper is Heaven's spiritual temple. The stars of the Big Dipper have the power to remove Shen Qi Jing blockages of all life, blessing your Tao journey beyond comprehension.

Figure 21. Big Dipper

南斗六星合一 Nan Dou Liu Xing He Yi

"Nan Dou" means *Southern Dipper* or *Little Dipper*. "Liu" means *six*. "Xing" means *star*. "Nan Dou Liu Xing" is formed by six stars of the constellation of Ursa Minor. It has a rough appearance of a dipper and so is paired with the Big Dipper in the northern sky. "Nan Dou Liu Xing He Yi" means *the Shen Qi Jing of the six stars in the Southern Dipper join as one*. The Southern Dipper is also a sacred Heaven's temple. Its power is also beyond imagination. You will receive great blessings from the Southern Dipper for your longevity and immortality journey. Chant more. Connect with the Southern Dipper. Benefits are unlimited.

五行合一 Wu Xing He Yi

"Wu Xing" refers to the Five Elements (figure 22). "Wu Xing He Yi" means *the Five Elements join as one*. The human body consists of Five Elements, including Wood, Fire, Earth, Metal, and Water. To join the Five Elements as one is a major step to reach Tao.

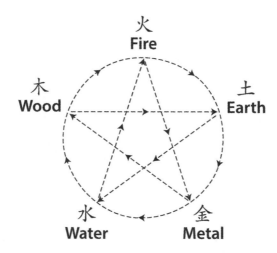

Figure 22. Five Elements

四象合一 Si Xiang He Yi

"Si Xiang" in figure 23 comes from *I Ching,* which refers to four forms produced by Yin and Yang: Lao Yang (老陽), Shao Yang (少陽), Lao Yin (老陰), Shao Yin (少陰). "Lao" means *old.* "Shao" means *young.* "Si Xiang He Yi" means *Lao Yang, Shao Yang, Lao Yin, and Shao Yin join as one.* To join Si Xiang as one is another key practice for reaching Tao.

Figure 23: Si Xiang

神氣精合一 Shen Qi Jing He Yi

"Shen" includes *soul, heart, and mind*. "Qi" means *energy*. "Jing" means *matter*. "Shen Qi Jing He Yi" is *to join soul, heart, mind, energy, and matter as one*. To join Shen Qi Jing as one is vital for reaching Tao.

陰陽合一 Yin Yang He Yi

Yin Yang theory comes from ancient Taoism, traditional Chinese medicine, and *I Ching*. Yin Yang theory and practice are universal laws that summarize everyone and everything in countless planets, stars, galaxies, and universes into two natures, which are Yin and Yang. Yin is the nature of water, which includes cold, flowing downward, calmness, and more. Yang is the nature of fire, which includes heat, flowing upward, excitement, and more. Yin Yang He Yi is Er Gui Yi (Two returns to One) in order to reach Tao.

人地天道中合一 Ren Di Tian Dao Zhong He Yi

"Ren" means *human being*. "Di" means *Mother Earth*. "Tian" means *Heaven*. Dao is the Source and Ultimate Creator. "Zhong" means *the core of life*. "Ren Di Tian Dao Zhong He Yi" means *join the cores of a human being, Mother Earth, Heaven, and Tao as one*. This is the final practice to reach Tao.

萬物合一 Wan Wu He Yi

"Wan" means *ten thousand*, which represents *countless*. "Wu" means *things*. "Wan Wu He Yi" means *join the Shen Qi Jing of all things, including countless planets, stars, galaxies, and universes as one*. All things can reach Tao.

Practice

Apply the Four Power Techniques:

1. Body Power

Sit or stand as in the practice in chapter one. Alternate sitting and standing positions in your practice for the best results.

2. Soul Power

Say *hello* to inner and outer souls:

> *Dear Tao Source,*
> *Dear Heaven and Mother Earth,*
> *Dear countless planets, stars, galaxies, and universes,*
> *Dear Bao Yuan Shou Yi Chang Shou Yong Sheng Dao Jing (Tao Classic of*
>   Longevity and Immortality),
> *I love you, honor you, and appreciate you.*
> *Please purify and remove my Shen Qi Jing blockages for healing, rejuvenation,*
>   *transformation of relationships and finances, increasing intelligence, and*
>   *flourishing in every aspect of my life.*
> *Please bless my spiritual practice of longevity and immortality to meld with*
>   *Tao.*
> *Please bless me to achieve my longevity and immortality journey.*
> *I am extremely honored and grateful.*
> *Thank you. Thank you. Thank you.*

3. Mind Power

Visualize a Tao Source Jin Dan (golden light ball) shining and rotating in your Zhong. The light is cleansing the Shen Qi Jing blockages in your Zhong and then getting brighter and brighter to grow to the size of the body (figure 24).

Figure 24. On the Way to Tao Practice

4. Sound Power

Chant silently or aloud the first seventeen stanzas of the *Tao Classic of Longevity and Immortality* (from chapters one through seventeen), and then repeat the seventeenth stanza (in **bold** below) ten or more times in each practice session. Remember to alternate silent and aloud (Yin Yang) chanting for best results.

Hei Hei Hei
Hei Jin Dan
Da Guang Ming
Da Guang Ming Jin Dan
Wo Zai Dao Guang Zhong
Dao Guang Zai Wo Zhong
Tong Ti Tou Ming

Heng Heng Heng
Heng Jin Dan
Da Gan En

Da Gan En Jin Dan
Dao Sheng De Yang
Zai Pei Ci Hui
Dao En Yong Cun

Hong Hong Hong
Hong Jin Dan
Da Qian Bei
Da Qian Bei Jin Dan
Rou Ruo Bu Zheng
Chi Xu Jing Jin
Shi Qian Bei
Die Wan Zhang

Ah Ah Ah
Ah Jin Dan
Da Ai
Da Ai Jin Dan
Wu Tiao Jian Ai
Rong Hua Zai Nan
Xin Qing Shen Ming

Xi Xi Xi
Xi Jin Dan
Da Fu Wu
Da Fu Wu Jin Dan
Shi Wei Gong Pu
Wu Si Feng Xian
Shang Cheng Fa Men

Yi Yi Yi
Yi Jin Dan
Da Kuan Shu
Da Kuan Shu Jin Dan
Wo Yuan Liang Ni
Ni Yuan Liang Wo
Xiang Ai Ping An He Xie

Weng Weng Weng
Weng Jin Dan
Da Yuan Man
Da Yuan Man Jin Dan
Ling Xin Nao Shen Yuan Man
Ren Di Tian Dao Shen Xian Ti
Fu Wu Xiu Lian Cai Ke Pan

You You You
You Jin Dan
Da Ci Bei
Da Ci Bei Jin Dan
Yuan Li Zeng Qiang
Fu Wu Zhong Sheng
Gong De Wu Liang

Ha Ha Ha
Ha Jin Dan
Da Chang Sheng
Da Chang Sheng Jin Dan
Dao Ci Ying Fu
Xing Shan Ji De
Dao Ye Chang Sheng

Yu Yu Yu
Yu Jin Dan
Da He Xie
Da He Xie Jin Dan
San Ren Tong Xin
Qi Li Duan Jin
Cheng Gong Mi Jue

Hei Heng Hong Ah Xi Yi Weng You
Hei Heng Hong Ah Xi Yi Weng You
Hei Heng Hong Ah Xi Yi Weng You
Hei Heng Hong Ah Xi Yi Weng You

You Weng Yi Xi Ah Hong Heng Hei
You Weng Yi Xi Ah Hong Heng Hei
You Weng Yi Xi Ah Hong Heng Hei
You Weng Yi Xi Ah Hong Heng Hei

Weng Hei Hong You
Weng Hei Hong You
Weng Hei Hong You
Weng Hei Hong You

Guang Gan Qian Ai Fu Kuan Yuan Ci
Guang Gan Qian Ai Fu Kuan Yuan Ci
Guang Gan Qian Ai Fu Kuan Yuan Ci
Guang Gan Qian Ai Fu Kuan Yuan Ci

Ci Yuan Kuan Fu Ai Qian Gan Guang
Ci Yuan Kuan Fu Ai Qian Gan Guang
Ci Yuan Kuan Fu Ai Qian Gan Guang
Ci Yuan Kuan Fu Ai Qian Gan Guang

Yuan Guang Qian Ci
Yuan Guang Qian Ci
Yuan Guang Qian Ci
Yuan Guang Qian Ci

Xu Xu Xu
Xu Jin Dan
Shu Gan Jin Dan

Ah Ah Ah
Ah Jin Dan
Yang Xin Jin Dan

Hu Hu Hu
Hu Jin Dan
Jian Pi Jin Dan

Si Si Si
Si Jin Dan
Xuan Fei Jin Dan

Chui Chui Chui
Chui Jin Dan
Zhuang Shen Jin Dan

Mu Bu Wang Shi
Er Bu Wang Ting
Kou Bu Wang Yan
Wai San Bao Bu Lou
Bu Shi An Shen Yu Xin
Bu Ting Xu Jing Yu Shen
Bu Yan Yun Qi Dan Tian
Nei San Bao Zi He
Wai Bu Lou
Nei Zi He
Tong Tian Da Di
Xiao Yao Dao Zhong

Bao Yuan Shou Yi
Yi Ji Shi Zhong
Kun Gong Ming Men Wei Lü Hui Yin Zu Gen Zhong
Ren Zhong Di Zhong
Tian Zhong Dao Zhong
Zhong Zhong Dao Kong Kong

Bao Yuan Shou Yi
Tian Yi Zhen Shui
Jin Jin Yu Ye
Yan Ru Dan Tian
Shen Qi Jing He Yi
Tian Di Ren He Yi
Jin Dan Xing Cheng

Bao Yuan Shou Yi
She Di Shang E
Yan Jin Bu Duan
Jin Dan Zhuang Da (1 time)

Bao Yuan Shou Yi
Shi Yuan Man He Yi
Jiu Jiu Gui Yi
Ba Gua He Yi
Bei Dou Qi Xing He Yi
Nan Dou Liu Xing He Yi
Wu Xing He Yi
Si Xiang He Yi
Shen Qi Jing He Yi
Yin Yang He Yi
Ren Di Tian Dao Zhong He Yi
Wan Wu He Yi (10 times)

Chapter Eighteen

Chinese	Pinyin	English Translation
道生一	Dao Sheng Yi	*Tao creates One*
一生二	Yi Sheng Er	*One creates Two*
二生三	Er Sheng San	*Two creates Three*
三生萬物	San Sheng Wan Wu	*Three creates all things*
萬物歸三	Wan Wu Gui San	*All things purify in order to return to and reach Three*
三歸二	San Gui Er	*Three purifies in order to return to and reach Two*
二歸一	Er Gui Yi	*Two purifies in order to return to and reach One*
一歸道	Yi Gui Dao	*One purifies in order to return to and reach Tao*

Sacred wisdom

道生一 Dao Sheng Yi
Tao creates One: Tao creates Hun Dun Yi Qi, Oneness.

一生二 Yi Sheng Er
One creates Two: Hun Dun Yi Qi, Oneness, creates Heaven and Earth.

二生三 Er Sheng San
Two creates Three: Heaven and Earth plus Hun Dun Yi Qi are Three.

三生萬物 San Sheng Wan Wu

Three creates all things: Hun Dun Yi Qi plus Heaven and Earth create all things.

萬物歸三 Wan Wu Gui San

All things purify and join their Shen Qi Jing together in order to return to the state of Three, which is the Shen Qi Jing level of Hun Dun Yi Qi plus Heaven and Earth.

三歸二 San Gui Er

Purify and join the Shen Qi Jing of the state of Three together in order to return to the state of Two (Heaven and Earth).

二歸一 Er Gui Yi

Purify and join the Shen Qi Jing of the state of Two together in order to return to the state of One, Hun Dun Yi Qi.

一歸道 Yi Gui Dao

Purify and join the Shen Qi Jing of One, Hun Dun Yi Qi, together in order to return to Tao.

This is Tao Normal Creation and Tao Reverse Creation (figure 1). This is the ultimate truth of the longevity, immortality, and Tao journey.

Practice

Apply the Four Power Techniques:

1. Body Power

Sit or stand as in the practice in chapter one. Alternate sitting and standing positions for the best results.

2. Soul Power

Say *hello* to inner and outer souls:

Dear Tao Source,

Dear Heaven and Mother Earth,

Dear countless planets, stars, galaxies, and universes,

Dear Bao Yuan Shou Yi Chang Shou Yong Sheng Dao Jing (Tao Classic of
    Longevity and Immortality),

I love you, honor you, and appreciate you.

*Please purify and remove my Shen Qi Jing blockages for healing, rejuvenation,
    transformation of relationships and finances, increasing intelligence, and
    flourishing in every aspect of my life.*

*Please bless my spiritual practice of longevity and immortality and my Tao
    Normal Creation and Tao Reverse Creation journey in order to meld with
    Tao.*

Please bless me to achieve my longevity and immortality journey.

I am extremely honored and grateful.

Thank you. Thank you. Thank you.

3. Mind Power

Visualize a Tao Source Jin Dan (golden light ball) shining and rotating from
your Zhong area to cover your entire body (figure 25) and then getting brighter
and brighter.

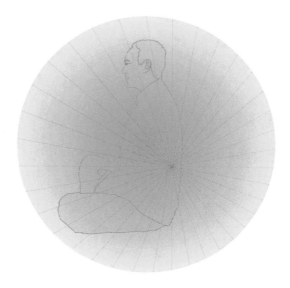

Figure 25. Meld with Tao practice

4. Sound Power

Chant silently or aloud the first eighteen stanzas of the *Tao Classic of Longevity and Immortality* (from chapters one through eighteen), and then repeat the eighteenth stanza (in **bold** below) ten or more times in each practice session. Alternate silent and aloud (Yin Yang) chanting for best results.

Hei Hei Hei
Hei Jin Dan
Da Guang Ming
Da Guang Ming Jin Dan
Wo Zai Dao Guang Zhong
Dao Guang Zai Wo Zhong
Tong Ti Tou Ming

Heng Heng Heng
Heng Jin Dan
Da Gan En

Da Gan En Jin Dan
Dao Sheng De Yang
Zai Pei Ci Hui
Dao En Yong Cun

Hong Hong Hong
Hong Jin Dan
Da Qian Bei
Da Qian Bei Jin Dan
Rou Ruo Bu Zheng
Chi Xu Jing Jin
Shi Qian Bei
Die Wan Zhang

Ah Ah Ah
Ah Jin Dan
Da Ai
Da Ai Jin Dan
Wu Tiao Jian Ai
Rong Hua Zai Nan
Xin Qing Shen Ming

Xi Xi Xi
Xi Jin Dan
Da Fu Wu
Da Fu Wu Jin Dan
Shi Wei Gong Pu
Wu Si Feng Xian
Shang Cheng Fa Men

Yi Yi Yi
Yi Jin Dan
Da Kuan Shu
Da Kuan Shu Jin Dan
Wo Yuan Liang Ni
Ni Yuan Liang Wo
Xiang Ai Ping An He Xie

Weng Weng Weng
Weng Jin Dan
Da Yuan Man
Da Yuan Man Jin Dan
Ling Xin Nao Shen Yuan Man
Ren Di Tian Dao Shen Xian Ti
Fu Wu Xiu Lian Cai Ke Pan

You You You
You Jin Dan
Da Ci Bei
Da Ci Bei Jin Dan
Yuan Li Zeng Qiang
Fu Wu Zhong Sheng
Gong De Wu Liang

Ha Ha Ha
Ha Jin Dan
Da Chang Sheng
Da Chang Sheng Jin Dan
Dao Ci Ying Fu
Xing Shan Ji De
Dao Ye Chang Sheng

Yu Yu Yu
Yu Jin Dan
Da He Xie
Da He Xie Jin Dan
San Ren Tong Xin
Qi Li Duan Jin
Cheng Gong Mi Jue

Hei Heng Hong Ah Xi Yi Weng You
Hei Heng Hong Ah Xi Yi Weng You
Hei Heng Hong Ah Xi Yi Weng You
Hei Heng Hong Ah Xi Yi Weng You

You Weng Yi Xi Ah Hong Heng Hei
You Weng Yi Xi Ah Hong Heng Hei
You Weng Yi Xi Ah Hong Heng Hei
You Weng Yi Xi Ah Hong Heng Hei

Weng Hei Hong You
Weng Hei Hong You
Weng Hei Hong You
Weng Hei Hong You

Guang Gan Qian Ai Fu Kuan Yuan Ci
Guang Gan Qian Ai Fu Kuan Yuan Ci
Guang Gan Qian Ai Fu Kuan Yuan Ci
Guang Gan Qian Ai Fu Kuan Yuan Ci

Ci Yuan Kuan Fu Ai Qian Gan Guang
Ci Yuan Kuan Fu Ai Qian Gan Guang
Ci Yuan Kuan Fu Ai Qian Gan Guang
Ci Yuan Kuan Fu Ai Qian Gan Guang

Yuan Guang Qian Ci
Yuan Guang Qian Ci
Yuan Guang Qian Ci
Yuan Guang Qian Ci

Xu Xu Xu
Xu Jin Dan
Shu Gan Jin Dan
Ah Ah Ah
Ah Jin Dan
Yang Xin Jin Dan

Hu Hu Hu
Hu Jin Dan
Jian Pi Jin Dan

Si Si Si
Si Jin Dan
Xuan Fei Jin Dan

Chui Chui Chui
Chui Jin Dan
Zhuang Shen Jin Dan

Mu Bu Wang Shi
Er Bu Wang Ting
Kou Bu Wang Yan
Wai San Bao Bu Lou
Bu Shi An Shen Yu Xin
Bu Ting Xu Jing Yu Shen
Bu Yan Yun Qi Dan Tian
Nei San Bao Zi He
Wai Bu Lou
Nei Zi He
Tong Tian Da Di
Xiao Yao Dao Zhong

Bao Yuan Shou Yi
Yi Ji Shi Zhong
Kun Gong Ming Men Wei Lü Hui Yin Zu Gen Zhong
Ren Zhong Di Zhong
Tian Zhong Dao Zhong
Zhong Zhong Dao Kong Kong

Bao Yuan Shou Yi
Tian Yi Zhen Shui
Jin Jin Yu Ye
Yan Ru Dan Tian
Shen Qi Jing He Yi
Tian Di Ren He Yi
Jin Dan Xing Cheng

Bao Yuan Shou Yi
She Di Shang E
Yan Jin Bu Duan
Jin Dan Zhuang Da

Bao Yuan Shou Yi
Shi Yuan Man He Yi
Jiu Jiu Gui Yi
Ba Gua He Yi
Bei Dou Qi Xing He Yi
Nan Dou Liu Xing He Yi
Wu Xing He Yi
Si Xiang He Yi
Shen Qi Jing He Yi
Yin Yang He Yi
Ren Di Tian Dao Zhong He Yi
Wan Wu He Yi (1 time)

Dao Sheng Yi
Yi Sheng Er
Er Sheng San
San Sheng Wan Wu
Wan Wu Gui San
San Gui Er
Er Gui Yi
Yi Gui Dao (10 times)

Chapter Nineteen

Chinese	Pinyin	English Translation
常誦此經不離口	Chang Song Ci Jing Bu Li Kou	*Continue to chant this sacred classic nonstop, silently or aloud*
常丹轉此經	Chang Dan Zhuan Ci Jing	*Frequently trace this sacred classic with your Dan*
金丹持續壯大	Jin Dan Chi Xu Zhuang Da	*Jin Dan continues to grow bigger and stronger*
待到金丹等身大	Dai Dao Jin Dan Deng Shen Da	*Wait until the Jin Dan grows to the size of the body*
長壽永生道修成	Chang Shou Yong Sheng Dao Xiu Cheng	*Longevity, immortality, and the Tao journey are accomplished*

Sacred wisdom

常誦此經不離口 Chang Song Ci Jing Bu Li Kou

"Chang" means *often, consistently*. "Song" means *to chant*. "Ci" means *this*. "Jing" refers to the *Tao Classic of Longevity and Immortality*. "Bu" means *not*. "Li" means *leave*. "Kou" means *mouth*. "Chang Song Ci Jing Bu Li Kou" means *chant this Tao Classic nonstop*. Do this to purify and remove Shen Qi Jing blockages in order to accomplish longevity and immortality. This is the ultimate purpose and goal of this Tao Classic.

常丹轉此經 Chang Dan Zhuan Ci Jing

"Chang" means *often, consistently*. "Dan" means *Lower Dan Tian*. "Zhuan" means *to trace*. "Ci" means *this*. "Jing" refers to the *Tao Classic of Longevity and Immortality*. "Chang Dan Zhuan Ci Jing" means *rotate your lower abdomen to trace the text of this Tao Classic consistently*. This practice can accelerate your purification and removal of Shen Qi Jing blockages in order to accomplish longevity and immortality.

金丹持續壯大 Jin Dan Chi Xu Zhuang Da

"Jin Dan" is the golden light ball. "Chi Xu" means *continuously*. "Zhuang Da" means *expanding strongly*. "Jin Dan Chi Xu Zhuang Da" means *the Jin Dan continues to grow bigger and stronger*.

待到金丹等身大 Dai Dao Jin Dan Deng Shen Da

"Dai Dao" means *wait until the time*. "Deng Shen Da" means *to equal the size of your body*. "Dai Dao Jin Dan Deng Shen Da" means *wait until your Jin Dan grows to the size of your body*.

長壽永生道修成 Chang Shou Yong Sheng Dao Xiu Cheng

"Chang Shou" means *longevity*. "Yong Sheng" means *immortality*. Here, "Dao" means *the immortality journey*. "Xiu" means *purify*. "Cheng" means *to succeed* or *complete*. "Chang Shou Yong Sheng Dao Xiu Cheng" means *longevity and immortality have been accomplished*.

Practice

Apply the Four Power Techniques:

1. Body Power

Sit or stand as in the practice in chapter one. Alternate sitting and standing positions for the best results.

2. Soul Power

Say *hello* to inner and outer souls:

> *Dear Tao Source,*
> *Dear Heaven and Mother Earth,*
> *Dear countless planets, stars, galaxies, and universes,*
> *Dear Bao Yuan Shou Yi Chang Shou Yong Sheng Dao Jing* (Tao Classic of
>   Longevity and Immortality),
> *I love you, honor you, and appreciate you.*
> *Please bless me to achieve longevity and immortality.*
> *I am extremely honored and grateful.*
> *Thank you. Thank you. Thank you.*

3. Mind Power

Visualize a Tao Source Jin Dan (golden light ball) shining and rotating in your whole body with bright Tao light (figure 26).

Figure 26. Immortality and Tao are reached

4. Sound Power

Chant silently or aloud all nineteen stanzas of the *Tao Classic of Longevity and Immortality* (from chapters one through nineteen), and then repeat the nineteenth stanza (in **bold** below) ten or more times in each practice session. Always alternate silent and aloud (Yin Yang) chanting.

Hei Hei Hei
Hei Jin Dan
Da Guang Ming
Da Guang Ming Jin Dan
Wo Zai Dao Guang Zhong
Dao Guang Zai Wo Zhong
Tong Ti Tou Ming

Heng Heng Heng
Heng Jin Dan
Da Gan En
Da Gan En Jin Dan
Dao Sheng De Yang
Zai Pei Ci Hui
Dao En Yong Cun

Hong Hong Hong
Hong Jin Dan
Da Qian Bei
Da Qian Bei Jin Dan
Rou Ruo Bu Zheng
Chi Xu Jing Jin
Shi Qian Bei
Die Wan Zhang

Ah Ah Ah
Ah Jin Dan
Da Ai
Da Ai Jin Dan
Wu Tiao Jian Ai

Rong Hua Zai Nan
Xin Qing Shen Ming

Xi Xi Xi
Xi Jin Dan
Da Fu Wu
Da Fu Wu Jin Dan
Shi Wei Gong Pu
Wu Si Feng Xian
Shang Cheng Fa Men

Yi Yi Yi
Yi Jin Dan
Da Kuan Shu
Da Kuan Shu Jin Dan
Wo Yuan Liang Ni
Ni Yuan Liang Wo
Xiang Ai Ping An He Xie

Weng Weng Weng
Weng Jin Dan
Da Yuan Man
Da Yuan Man Jin Dan
Ling Xin Nao Shen Yuan Man
Ren Di Tian Dao Shen Xian Ti
Fu Wu Xiu Lian Cai Ke Pan

You You You
You Jin Dan
Da Ci Bei
Da Ci Bei Jin Dan
Yuan Li Zeng Qiang
Fu Wu Zhong Sheng
Gong De Wu Liang

Ha Ha Ha
Ha Jin Dan

Da Chang Sheng
Da Chang Sheng Jin Dan
Dao Ci Ying Fu
Xing Shan Ji De
Dao Ye Chang Sheng

Yu Yu Yu
Yu Jin Dan
Da He Xie
Da He Xie Jin Dan
San Ren Tong Xin
Qi Li Duan Jin
Cheng Gong Mi Jue

Hei Heng Hong Ah Xi Yi Weng You
Hei Heng Hong Ah Xi Yi Weng You
Hei Heng Hong Ah Xi Yi Weng You
Hei Heng Hong Ah Xi Yi Weng You

You Weng Yi Xi Ah Hong Heng Hei
You Weng Yi Xi Ah Hong Heng Hei
You Weng Yi Xi Ah Hong Heng Hei
You Weng Yi Xi Ah Hong Heng Hei

Weng Hei Hong You
Weng Hei Hong You
Weng Hei Hong You
Weng Hei Hong You

Guang Gan Qian Ai Fu Kuan Yuan Ci
Guang Gan Qian Ai Fu Kuan Yuan Ci
Guang Gan Qian Ai Fu Kuan Yuan Ci
Guang Gan Qian Ai Fu Kuan Yuan Ci

Ci Yuan Kuan Fu Ai Qian Gan Guang
Ci Yuan Kuan Fu Ai Qian Gan Guang
Ci Yuan Kuan Fu Ai Qian Gan Guang

Ci Yuan Kuan Fu Ai Qian Gan Guang

Yuan Guang Qian Ci
Yuan Guang Qian Ci
Yuan Guang Qian Ci
Yuan Guang Qian Ci

Xu Xu Xu
Xu Jin Dan
Shu Gan Jin Dan

Ah Ah Ah
Ah Jin Dan
Yang Xin Jin Dan

Hu Hu Hu
Hu Jin Dan
Jian Pi Jin Dan

Si Si Si
Si Jin Dan
Xuan Fei Jin Dan

Chui Chui Chui
Chui Jin Dan
Zhuang Shen Jin Dan

Mu Bu Wang Shi
Er Bu Wang Ting
Kou Bu Wang Yan
Wai San Bao Bu Lou
Bu Shi An Shen Yu Xin
Bu Ting Xu Jing Yu Shen
Bu Yan Yun Qi Dan Tian
Nei San Bao Zi He
Wai Bu Lou
Nei Zi He

Tong Tian Da Di
Xiao Yao Dao Zhong

Bao Yuan Shou Yi
Yi Ji Shi Zhong
Kun Gong Ming Men Wei Lü Hui Yin Zu Gen Zhong
Ren Zhong Di Zhong
Tian Zhong Dao Zhong
Zhong Zhong Dao Kong Kong

Bao Yuan Shou Yi
Tian Yi Zhen Shui
Jin Jin Yu Ye
Yan Ru Dan Tian
Shen Qi Jing He Yi
Tian Di Ren He Yi
Jin Dan Xing Cheng

Bao Yuan Shou Yi
She Di Shang E
Yan Jin Bu Duan
Jin Dan Zhuang Da

Bao Yuan Shou Yi
Shi Yuan Man He Yi
Jiu Jiu Gui Yi
Ba Gua He Yi
Bei Dou Qi Xing He Yi
Nan Dou Liu Xing He Yi
Wu Xing He Yi
Si Xiang He Yi
Shen Qi Jing He Yi
Yin Yang He Yi
Ren Di Tian Dao Zhong He Yi
Wan Wu He Yi

Dao Sheng Yi
Yi Sheng Er
Er Sheng San
San Sheng Wan Wu
Wan Wu Gui San
San Gui Er
Er Gui Yi
Yi Gui Dao (1 time)

Chang Song Ci Jing Bu Li Kou
Chang Dan Zhuan Ci Jing
Jin Dan Chi Xu Zhuang Da
Dai Dao Jin Dan Deng Shen Da
Chang Shou Yong Sheng Dao Xiu Cheng
Chang Shou Yong Sheng Dao Xiu Cheng
Chang Shou Yong Sheng Dao Xiu Cheng (10 times)

I am extremely honored to share the highest sacred Tao wisdom and practice of longevity and immortality within this new Tao Classic. Billions of people in history have searched for the sacred way to reach longevity and immortality. They have searched and searched for thousands or millions of lifetimes. The true secrets were hidden and very difficult to discover.

In this book, the sacred wisdom and practice to reach longevity and immortality are presented in this sacred Tao Classic. If you have read the book up to this moment, congratulations! What I would like to suggest to every reader is to master the wisdom. The most important thing is to put this wisdom into practice. If you have read the book and done the practices, greatest congratulations!

In ancient sacred teaching there is one sacred sentence:

To chant Tao is to reach Tao.

This new Tao Classic carries the Shen Qi Jing of Tao. It could remove your Shen Qi Jing blockages for your Tao journey to reach Tao. Please study and practice this Tao Classic as much as possible. Practice consistently. Practice with me singing this Tao Classic accompanied by beautiful Tao music. My singing will bless and empower your longevity, immortality, and Tao journey every time you listen to it.[8] You *can* achieve longevity and immortality.

Finally, I will bless you with the last sentence of this Tao Classic for your longevity, immortality, and Tao journey.

Chang Shou Yong Sheng Dao Xiu Cheng!
Chang Shou Yong Sheng Dao Xiu Cheng!
Chang Shou Yong Sheng Dao Xiu Cheng!

---

8 You can purchase the recording here: http://webshop-ca.drsha.com/melding-with-tao-classic-of -longevity-immortality.html.

Appendix

Bao Yuan Shou Yi Chang Shou Yong Sheng Dao Jing

抱元守一長壽永生道經
Tao Classic of Longevity and Immortality

	Chinese	Pinyin	English Translation
1	嘿嘿嘿	Hei Hei Hei	*Hei Hei Hei*
2	嘿金丹	Hei Jin Dan	*Hei Golden Light Ball*
3	大光明	Da Guang Ming	*Greatest Light*
4	大光明金丹	Da Guang Ming Jin Dan	*Greatest Light Golden Light Ball*
5	我在道光中	Wo Zai Dao Guang Zhong	*I am inside Tao Source light*
6	道光在我中	Dao Guang Zai Wo Zhong	*Tao Source light is inside me*
7	通體透明	Tong Ti Tou Ming	*Whole body is bright and transparent in the Tao light*
8	哼哼哼	Heng Heng Heng	*Heng Heng Heng*
9	哼金丹	Heng Jin Dan	*Heng Golden Light Ball*
10	大感恩	Da Gan En	*Greatest Gratitude*
11	大感恩金丹	Da Gan En Jin Dan	*Greatest Gratitude Golden Light Ball*
12	道生德養	Dao Sheng De Yang	*Tao Source creates and De nourishes everyone and everything*
13	栽培賜慧	Zai Pei Ci Hui	*Tao Source cultivates and bestows wisdom*

	Chinese	Pinyin	English Translation
14	道恩永存	Dao En Yong Cun	*The glory of Tao Source is eternal*
15	哄哄哄	Hong Hong Hong	*Hong Hong Hong*
16	哄金丹	Hong Jin Dan	*Hong Golden Light Ball*
17	大謙卑	Da Qian Bei	*Greatest Humility*
18	大謙卑金丹	Da Qian Bei Jin Dan	*Greatest Humility Golden Light Ball*
19	柔弱不爭	Rou Ruo Bu Zheng	*Be gentle and soft, do not compete*
20	持續精進	Chi Xu Jing Jin	*Improve persistently*
21	失謙卑	Shi Qian Bei	*Lose humility*
22	跌萬丈	Die Wan Zhang	*Fail tremendously in every aspect of life, just like falling into a deep cavern*
23	啊啊啊	Ah Ah Ah	*Ah Ah Ah*
24	啊金丹	Ah Jin Dan	*Ah Golden Light Ball*
25	大愛	Da Ai	*Greatest Love*
26	大愛金丹	Da Ai Jin Dan	*Greatest Love Golden Light Ball*
27	無條件愛	Wu Tiao Jian Ai	*Unconditional love*
28	融化災難	Rong Hua Zai Nan	*Melts all blockages*
29	心清神明	Xin Qing Shen Ming	*Purifies heart and enlightens soul, heart, and mind*
30	唏唏唏	Xi Xi Xi	*Xi Xi Xi*
31	唏金丹	Xi Jin Dan	*Xi Golden Light Ball*
32	大服務	Da Fu Wu	*Greatest Service*
33	大服務金丹	Da Fu Wu Jin Dan	*Greatest Service Golden Light Ball*

	Chinese	Pinyin	English Translation
34	誓為公僕	Shi Wei Gong Pu	*Vow to be a servant of humanity, Mother Earth, and Heaven*
35	無私奉獻	Wu Si Feng Xian	*Serve selflessly*
36	上乘法門	Shang Cheng Fa Men	*The highest way to reach Tao Source*
37	噫噫噫	Yi Yi Yi	*Yi Yi Yi*
38	噫金丹	Yi Jin Dan	*Yi Golden Light Ball*
39	大寬恕	Da Kuan Shu	*Greatest Forgiveness*
40	大寬恕金丹	Da Kuan Shu Jin Dan	*Greatest Forgiveness Golden Light Ball*
41	我原諒你	Wo Yuan Liang Ni	*I forgive you*
42	你原諒我	Ni Yuan Liang Wo	*You forgive me*
43	相愛平安和諧	Xiang Ai Ping An He Xie	*Love, peace, and harmony*
44	嗡嗡嗡	Weng Weng Weng	*Weng Weng Weng*
45	嗡金丹	Weng Jin Dan	*Weng Golden Light Ball*
46	大圓滿	Da Yuan Man	*Greatest Enlightenment*
47	大圓滿金丹	Da Yuan Man Jin Dan	*Greatest Enlightenment Golden Light Ball*
48	靈心腦身圓滿	Ling Xin Nao Shen Yuan Man	*Soul heart mind body enlightenment*
49	人地天道神仙梯	Ren Di Tian Dao Shen Xian Ti	*Human saint, Mother Earth saint, Heaven saint, and Tao Source saint are enlightenment stairs*
50	服務修煉才可攀	Fu Wu Xiu Lian Cai Ke Pan	*Only through service and purification can one climb the stairs and reach all layers of enlightenment*

	Chinese	Pinyin	English Translation
51	呦呦呦	You You You	*You You You*
52	呦金丹	You Jin Dan	*You Golden Light Ball*
53	大慈悲	Da Ci Bei	*Greatest Compassion*
54	大慈悲金丹	Da Ci Bei Jin Dan	*Greatest Compassion Golden Light Ball*
55	願力增強	Yuan Li Zeng Qiang	*Increase willpower*
56	服務眾生	Fu Wu Zhong Sheng	*Serve humanity*
57	功德無量	Gong De Wu Liang	*Virtue will be immeasurable*
58	哈哈哈	Ha Ha Ha	*Ha Ha Ha*
59	哈金丹	Ha Jin Dan	*Ha Golden Light Ball*
60	大昌盛	Da Chang Sheng	*Greatest Flourishing*
61	大昌盛金丹	Da Chang Sheng Jin Dan	*Greatest Flourishing Golden Light Ball*
62	道賜盈福	Dao Ci Ying Fu	*Tao Source bestows huge fortune, including Tao wisdom, health, longevity, happiness, good relationships, prosperity, luck, success, and more to all aspects of life*
63	行善積德	Xing Shan Ji De	*Be kind to accumulate virtue*
64	道業昌盛	Dao Ye Chang Sheng	*Tao career flourishes*
65	吁吁吁	Yu Yu Yu	*Yu Yu Yu*
66	吁金丹	Yu Jin Dan	*Yu Golden Light Ball*
67	大和諧	Da He Xie	*Greatest Harmony*

	Chinese	Pinyin	English Translation
68	大和諧金丹	Da He Xie Jin Dan	*Greatest Harmony Golden Light Ball*
69	三人同心	San Ren Tong Xin	*Three people join hearts together*
70	齊力斷金	Qi Li Duan Jin	*Their strength can cut through gold*
71	成功秘訣	Cheng Gong Mi Jue	*This is the Tao secret of success*
72	嘿哼哄啊唏噎嗡呦	Hei Heng Hong Ah Xi Yi Weng You	*Qi Channel for healing*
73	呦嗡噎唏啊哄哼嘿	You Weng Yi Xi Ah Hong Heng Hei	*Jing Channel for rejuvenation and longevity*
74	嗡嘿哄呦	Weng Hei Hong You	*Shen Channel for immortality*
75	光感謙愛服寬圓慈	Guang Gan Qian Ai Fu Kuan Yuan Ci	*Tao Source Eight Da Natures Qi Channel for healing*
76	慈圓寬服愛謙感光	Ci Yuan Kuan Fu Ai Qian Gan Guang	*Tao Source Eight Da Natures Jing Channel for rejuvenation and longevity*
77	圓光謙慈	Yuan Guang Qian Ci	*Tao Source Four Da Natures Shen Channel for immortality*
78	噓噓噓	Xu Xu Xu	*Xu Xu Xu*
79	噓金丹	Xu Jin Dan	*Xu Golden Light Ball*
80	舒肝金丹	Shu Gan Jin Dan	*Smooth Liver Function Golden Light Ball*
81	啊啊啊	Ah Ah Ah	*Ah Ah Ah*
82	啊金丹	Ah Jin Dan	*Ah Golden Light Ball*

	Chinese	Pinyin	English Translation
83	養心金丹	Yang Xin Jin Dan	*Nourish Heart Golden Light Ball*
84	呼呼呼	Hu Hu Hu	*Hu Hu Hu*
85	呼金丹	Hu Jin Dan	*Hu Golden Light Ball*
86	健脾金丹	Jian Pi Jin Dan	*Strengthen Spleen Golden Light Ball*
87	呬呬呬	Si Si Si	*Si Si Si*
88	呬金丹	Si Jin Dan	*Si Golden Light Ball*
89	宣肺金丹	Xuan Fei Jin Dan	*Facilitate Lung Function Golden Light Ball*
90	吹吹吹	Chui Chui Chui	*Chui Chui Chui*
91	吹金丹	Chui Jin Dan	*Chui Golden Light Ball*
92	壯腎金丹	Zhuang Shen Jin Dan	*Reinforce Kidneys Golden Light Ball*
93	目不妄視	Mu Bu Wang Shi	*Do not look at impure things*
94	耳不妄聽	Er Bu Wang Ting	*Do not listen to impure sounds*
95	口不妄言	Kou Bu Wang Yan	*Do not speak impure words*
96	外三寶不漏	Wai San Bao Bu Lou	*Three external treasures (eyes, ears, nose) do not leak*
97	不視安神於心	Bu Shi An Shen Yu Xin	*Do not look at impure things to keep your Shen (soul, heart, mind) pure and peaceful in your heart*

	Chinese	Pinyin	English Translation
98	不聽蓄精於腎	Bu Ting Xu Jing Yu Shen	*Do not listen to impure sounds to increase the pure matter essence in your kidneys*
99	不言孕氣丹田	Bu Yan Yun Qi Dan Tian	*Do not speak impure words to increase the pure energy in your Lower Dan Tian*
100	內三寶自合	Nei San Bao Zi He	*Three internal treasures (Jing [matter], Qi [energy], Shen [soul, heart, mind]) join together as one*
101	外不漏	Wai Bu Lou	*Three external treasures (eyes, ears, mouth) do not leak*
102	內自合	Nei Zi He	*Three internal treasures (Shen Qi Jing) join as one*
103	通天達地	Tong Tian Da Di	*Connect through Heaven and Mother Earth as one*
104	逍遙道中	Xiao Yao Dao Zhong	*Meld with Tao and flow freely in the Tao*
105	抱元守一	Bao Yuan Shou Yi	*Hold Tao, focus on One, which is Tao*
106	一即是中	Yi Ji Shi Zhong	*One is Zhong, the core of one's life, which is Tao*
107	坤宮命門尾閭會陰足跟中	Kun Gong Ming Men Wei Lü Hui Yin Zu Gen Zhong	*Kun Temple, Ming Men, sacrum, perineum, and heels of feet Zhong*
108	人中地中	Ren Zhong Di Zhong	*Zhong includes Human Zhong, Mother Earth Zhong*

	Chinese	**Pinyin**	**English Translation**
109	天中道中	Tian Zhong Dao Zhong	*Zhong also includes Heaven Zhong, Tao Zhong*
110	中中道空空	Zhong Zhong Dao Kong Kong	*Core, core, Tao, emptiness, emptiness, which are all Tao Source*
111	抱元守一	Bao Yuan Shou Yi	*Hold Tao, focus on One, which is Tao*
112	天一真水	Tian Yi Zhen Shui	*Heaven's Oneness sacred nectar*
113	金津玉液	Jin Jin Yu Ye	*Earth's sacred golden and jade nectar*
114	咽入丹田	Yan Ru Dan Tian	*Swallow Heaven's and Earth's sacred nectars into the Lower Dan Tian*
115	神氣精合一	Shen Qi Jing He Yi	*Soul, heart, mind, energy, matter join as one*
116	天地人合一	Tian Di Ren He Yi	*Heaven's, Mother Earth's, human's soul, heart, mind, energy, matter join as one*
117	金丹形成	Jin Dan Xing Cheng	*Jin Dan is formed*
118	抱元守一	Bao Yuan Shou Yi	*Hold Tao, focus on One, which is Tao*
119	舌抵上顎	She Di Shang E	*Tip of tongue touches palate*
120	咽津不斷	Yan Jin Bu Duan	*Swallow Heaven's and Earth's sacred nectars constantly*
121	金丹壯大	Jin Dan Zhuang Da	*Jin Dan grows bigger and stronger*

	Chinese	Pinyin	English Translation
122	抱元守一	Bao Yuan Shou Yi	*Hold Tao, focus on One, which is Tao*
123	十圓滿合一	Shi Yuan Man He Yi	*Number ten indicates soul, heart, mind, and body enlightenments join as one*
124	九九歸一	Jiu Jiu Gui Yi	*Number nine indicates all kinds of spiritual belief systems join as one*
125	八卦合一	Ba Gua He Yi	*Number eight indicates I Ching Ba Gua, including Heaven, Earth, thunder, wind, water, fire, mountain, and lake, join as one*
126	北斗七星合一	Bei Dou Qi Xing He Yi	*Number seven indicates the seven stars in the Big Dipper join as one*
127	南斗六星合一	Nan Dou Liu Xing He Yi	*Number six indicates the six stars in the Southern Dipper join as one*
128	五行合一	Wu Xing He Yi	*Number five indicates Five Elements join as one*
129	四象合一	Si Xiang He Yi	*Number four indicates different conditions of Yin and Yang, including Young Yin, Old Yin, Young Yang, and Old Yang, join as one*
130	神氣精合一	Shen Qi Jing He Yi	*Soul, heart, mind, energy, matter join as one*
131	陰陽合一	Yin Yang He Yi	*Yin Yang join as one*
132	人地天道中合一	Ren Di Tian Dao Zhong He Yi	*Zhong of human being, Mother Earth, Heaven, and Tao join as one*

	Chinese	Pinyin	English Translation
133	萬物合一	Wan Wu He Yi	*Countless planets, stars, galaxies, and universes, human beings, and all things join as one*
134	道生一	Dao Sheng Yi	*Tao creates One*
135	一生二	Yi Sheng Er	*One creates Two*
136	二生三	Er Sheng San	*Two creates Three*
137	三生萬物	San Sheng Wan Wu	*Three creates all things*
138	萬物歸三	Wan Wu Gui San	*All things purify in order to return to and reach Three*
139	三歸二	San Gui Er	*Three purifies in order to return to and reach Two*
140	二歸一	Er Gui Yi	*Two purifies in order to return to and reach One*
141	一歸道	Yi Gui Dao	*One purifies in order to return to and reach Tao*
142	常誦此經不離口	Chang Song Ci Jing Bu Li Kou	*Continue to chant this sacred classic nonstop, silently or aloud*
143	常丹轉此經	Chang Dan Zhuan Ci Jing	*Lower Dan Tian traces this sacred classic frequently*
144	金丹持續壯大	Jin Dan Chi Xu Zhuang Da	*Jin Dan continues to grow bigger and stronger*
145	待到金丹等身大	Dai Dao Jin Dan Deng Shen Da	*Wait until the Jin Dan grows to the size of the body*
146	長壽永生道修成	Chang Shou Yong Sheng Dao Xiu Cheng	*Longevity, immortality, and the Tao journey have been accomplished*

About the Author

DR. AND MASTER Zhi Gang Sha is a world-renowned healer, Tao Grandmaster, philanthropist, humanitarian, and creator of Tao Calligraphy. He is the founder of Soul Mind Body Medicine™ and an eleven-time *New York Times* bestselling author. An MD in China and a doctor of traditional Chinese medicine in China and Canada, Master Sha is the founder of Tao Academy and the Love Peace Harmony Foundation™, which is dedicated to helping families worldwide create happier and healthier lives. A grandmaster of many ancient disciplines, including tai chi, qigong, kung fu, feng shui, and the *I Ching*, Master Sha was named Qigong Master of the Year at the Fifth World Congress on Qigong. In 2006, he was honored with the prestigious Martin Luther King, Jr. Commemorative Commission Award for his humanitarian efforts, and in 2016 Master Sha received rare and prestigious appointments as Shu Fa Jia (National Chinese Calligrapher Master) and Yan Jiu Yuan (Honorable Researcher Professor), the highest titles a Chinese calligrapher can receive, by the State Ethnic of Academy of Painting in China.